T0336614

Mycotoxins in Aquaculture

Mycotoxins in Aquaculture

Rui Gonçalves
Michele Muccio

Edited by

Anneliese Mueller

BIOMIN Edition

www.biomin.net
www.erber-group.net

A CIP catalogue record for this book is available from the British Library.

ISBN: 9781789180596

Disclaimer
Every reasonable effort has been made to ensure that the material in this book is true, correct, complete and appropriate at the time of writing. Nevertheless the editors and the authors do not accept responsibility for any omission or error, or for any injury, damage, loss or financial consequences arising from the use of the book.

Published by:
5m Publishing Ltd,
Benchmark House,
8 Smithy Wood Drive,
Sheffield, S35 1QN, UK
Tel: +44 (0) 1234 81 81 80
www.5mpublishing.com

Printed by PODWW, UK

Contents

List of acronyms

Afla	aflatoxins
Afla B_1	aflatoxin B_1
Afla B_2	aflatoxin B_2
Afla G_1	aflatoxin G_1
Afla G_2	aflatoxin G_2
Afla M_1	aflatoxin M_1
Afla P_1	aflatoxin P_1
Afla M_2	aflatoxin M_2
ALF	antilypopolisaccharide factor
Afla Q_1	aflatoxin Q_1
ALP	alkaline phosphatase
ALT	alanine aminotransferase
APC	antigen presenting cells
AST	aspartate transaminase
ATA	alimentary toxic aleukia
CD	cluster of differentiation
CerS	ceramide synthase
CGM	corn gluten meal
CSM	cottonseed meal
D3G	deoxynivalenol-3-glucoside
DAS	diacetoxyscirpenol
DDGS	dried distiller's grains with solubles
DOM-1	de-epoxy-deoxynivalenol
DON	deoxynivalenol
EFSA	European Food Safety Authority
ELEM	equine leukoencephalomalacia
ELISA	enzyme-linked immunosorbent assay
Ergots	ergot alkaloids
EU	European Union
FB_1	fumonisin B_1
FB_2	fumonisin B_2
FB_3	fumonisin B_3
FCR	feed conversion rate
FDA	Food and Drug Administration
FUM	fumonisins
GALT	gut-associated lymphoid tissue
GCs	granular cells
GIT	gastrointestinal tract
GST	glutathione S-transferase
HC	hyaline cells
HCK	hematopoietic cell kinase

HFB_1	hydrolyzed FB_1
HPLC	high performance liquid chromatography
HPT	hematopoietic tissues
HSI	hepatosomatic index
HT-2	HT-2 toxin
IARC	International Agency for Research on Cancer
INF-γ	interferon gamma
iNOS	inducible nitric oxide synthase
LC-MS	liquid chromatography-mass spectrometry
LO	lymphoid organ
LPS	lipopolysaccharide
MALT	mucosa-associated lymphoid tissue
MAPKs	mitogen-activated protein kinases
MS	mass spectrometry
MTV	*Trichosporon mycotoxinivorans*
NIV	nivalenol
NK	natural killer
NO	nitric oxide
NTB	nitroblue terazolium
OTA	ochratoxin A
OTB	ochratoxin B
OTHQ	hydroxyl quinone ochratoxin
OTQ	quinone ochratoxin
OTα	ochratoxin alpha
PAMP(s)	pathogen-associated molecular pattern(s)
PKR	RNA-activated protein kinase
OP-OTA	ochratoxin A open lactone
PPE	porcine pulmonary edema
proPO	prophenoloxidase
PRR(s)	pattern recognition receptor(s)
R/CM	rapeseed/canola meal
RB	rice bran
ROI	reactive oxygen intermediates
ROS	reactive oxygen species
Sa	sphinganine
SBM	soybean meal
SGCs	semigranular cells
So	sphingosine
SPT	serine palmitoyltransferase
T-2	T-2 toxin
TF	transcription factor
WB	wheat bran
WH	wheat
ZEN	zearalenone

List of figures and tables

Figures

Tables

Acknowledgement

We are particularly grateful for the assistance given by Dr. Christiane Gruber-Dorninger, who did a scientific review of all chapters and contributed with her critical view and constructive comments.

We would like to thank all people at BIOMIN supporting us and contributing to this book.

Introduction

With the current trend of fishmeal replacement in aquafeed, the issue of antinutrients contained in plant-based materials is getting of bigger concern for productivity. Mycotoxins are known antinutrients; however, their role in aquaculture has still to be fully elucidated. Generally, the industry tends to look at mycotoxins rather skeptically as they were never of big concern, but the interest of the scientific community towards these toxic metabolites is raising and a greater number of studies was published in recent times. In the last couple of years, we presented more insights into mycotoxin research in aquaculture at conferences, and could interestingly notice that the industry was open to learn more about this rather novel topic.

When we sat together and decided to write this book, we tried to imagine what would be useful for fish or shrimp producers and how they could practically use the information. We decided not only to provide a simple literature review on the toxicological effects on single species, but we wanted to create something that could help producers in their daily challenges and, at the same time, contribute to the scientific community by offering a new tool to whoever is approaching aquaculture in the era of finite resources.

Therefore, we decided to approach mycotoxins from the basis, providing general information to get the reader confident with the topic and then to have a look into the potential implications for the production chain. We also included our experience matured from terrestrial animals to potentially identify the targets of these antinutrients in aquatic species.

We do believe we have done our best to provide a comprehensive overview using all the published work, and a bit of our internal expertise as well. We had access to the work done by worldwide mycotoxin experts, many of them working at the IFA Institute in Tulln, Austria – or how we like to call it, the Silicon Valley of mycotoxins! We take this opportunity to thank our readers once again for having chosen our material and we hope to provide an engaging reading experience, being confident to see more aqua mycotoxin experts in the future.

Sincerely yours

Rui A. Gonçalves, Michele Muccio and Anneliese Müller

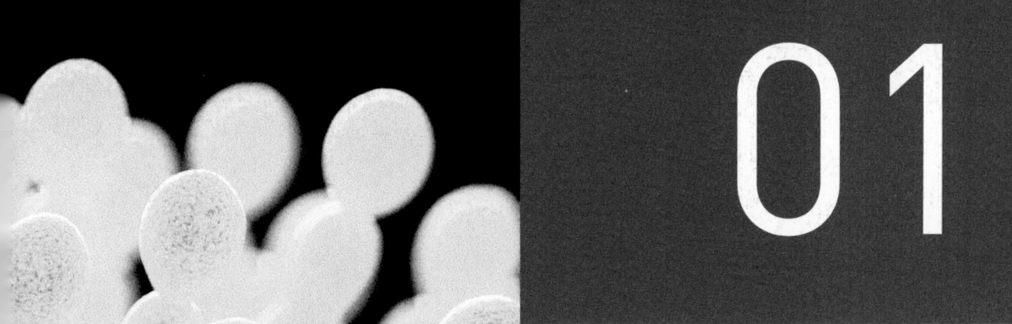

01

Mycotoxins – general concepts

Karin NAEHRER

Edited by Anneliese Mueller

1. Mycotoxins – general concepts

Karin NAEHRER

Edited by Anneliese Mueller

1.1 What are mycotoxins?

Mycotoxins are naturally occurring secondary metabolites produced by certain molds/fungi. Whereas primary metabolites (e.g. amino acids, sugars, etc.) are essential for the survival of organisms, the benefit of many secondary metabolites, such as mycotoxins, is not fully understood. Mycotoxins are chemical compounds of low molecular weight and low immunogenic capacity (Mallmann and Dilkin, 2007). There are some secondary metabolites of fungal origin with medicinal or industrial applications, for example, penicillin, but many exert detrimental effects on human and animal health, and animal productivity. Unfortunately, mycotoxins are known to contaminate crops and consequently animal feeds and animal products. In addition to the negative impacts on health, their presence can lead to significant economic losses. Crops with large amounts of mycotoxins often have to be destroyed (Vila-Donat *et al.*, 2018). The mycotoxins that cause the biggest economic impacts on animal production are aflatoxins (Afla), trichothecenes (namely deoxynivalenol [DON] and T-2 toxin [T-2]), zearalenone (ZEN), ochratoxin A (OTA), fumonisins (FUM) and ergot alkaloids (Ergots). The focus of this book will be on these mycotoxins.

1.2 Mycotoxin-producing fungi

In general, the process of mycotoxin production by fungi is not well understood. Fungi, just like any living organism need nutrients to survive and they might have to compete for plant nutrients with the plant itself as well as with other microorganisms. Thus, mycotoxins may allow the fungus to prevail in adverse conditions by conferring a competitive advantage over other organisms (Rankin and Grau, 2002). Some insights on the effect of mycotoxins on the host plant have been gained for the mycotoxin DON: Its production has been found to prevent the formation of a thick cell wall, which could help the plant to avoid fungal infection. Furthermore, its production has been reported to be induced as a response to host defenses. Deoxynivalenol has also been observed to help the fungi to infect further until now healthy plant parts (Khaneghah *et al.*, 2018). Unfortunately, even if we cannot entirely explain the reasons for their existence, mycotoxins are produced during different stages of food and feed production and are a serious problem worldwide.

Conditions suitable for fungal growth can occur at all times during crop growth, harvest and storage. Fungal species can be roughly categorized as field molds, which infect crops as parasites and storage fungi, which grow in feedstuffs stored under sub-optimal conditions.

Field fungi such as *Fusarium* spp. generally require higher moisture levels (> 0.9 water activity) to grow and produce mycotoxins. Therefore, they mainly infect seeds and plants in the field. Storage fungi such as *Aspergillus* spp. and *Penicillium* spp. require lower water activity and are thus more prominent after harvest and during storage (see Chapter 1.3).

Infection by *Claviceps* spp. and *Neotyphodium* spp. occurs only in the field. Claviceps are plant pathogens that replace plant structures such as grain kernels with hardened fungal tissues called ergots or sclerotia (Tudzynski *et al.*, 2001). Sclerotia are a protection mechanism of the fungus allowing it to survive adverse environmental conditions. The sclerotia subsequently produce conidiospores and sugar containing secretions to attract insects and favor dispersion of spores to new hosts. In autumn, the sclerotia fall to the ground and overwinter until the following spring. Infection is usually favored by a cold winter followed by a wet spring. The fungus uses the nutrients from the plant for the development of the sclerotia and the production of Ergots. Sclerotia often contain a broad range of Ergots. The Claviceps genus, mainly *Claviceps purpurea*, parasitizes more than 600 plant species including some economically important cereal grains such as rye, wheat, barley, millet and oats (Strickland *et al.*, 2011). In addition, ergot contamination in sorghum due to *Claviceps africana* has been discovered: here the Claviceps spores germinate and grow into the unfertilized seed producing a sclerotium (Krska and Crews, 2008).

Ergot alkaloids cause a disease known as ergotism, which was one of the first recognized mycotoxicoses (CAST, 2003; Flieger *et al.*, 1997). The alkaloid pattern and individual alkaloid content of sclerotia vary largely according to fungal strain, host plant, differences in the maturity of the sclerotia, geographical regions and weather conditions (European Food Safety Authority [EFSA], 2011a).

In Table 1.1 the most important mycotoxin-producing fungi and their mycotoxins are listed. Some substances listed in Table 1.1 are so called "emerging mycotoxins", for example, moniliformin (see Chapter 1.6).

***Table 1.1** – The most important mycotoxins and their producers*
Source: www.mycotoxins.info

Major classes of mycotoxin-producing fungi	Fungi species	Mycotoxins
Aspergillus	*A. flavus* *A. parasiticus* *A. nomius* *A. pseudotamarii*	Aflatoxin (B_1, B_2, G_1, G_2)
	A. ochraceus	Ochratoxin (Ochratoxin A)
	A. clavatus *A. terreus*	Patulin
	A. flavus *A. versicolor*	Cyclopiazonic acid
Claviceps	*C. purpurea* *C. fusiformis* *C. paspali* *C. africana*	**Ergot alkaloids:** Clavines (Agroclavine) Lysergic acids Lysergic acid amids (Ergin) Ergopeptines (Ergotamine, Ergovaline)

Fusarium	*F. verticillioides* *(syn. F. moniliforme)* *F. proliferatum*	Fumonisin (B_1, B_2, B_3) Fusaric Acid Moniliformin
	F. graminearum *F. avenaceum* *F. culmorum* *F. poae* *F. equiseti* *F. crookwellense* *F. acuminatum* *F. sambucinum* *F. sporotrichioides*	**Type-A Trichothecenes** T-2 toxin, HT-2 toxin, diacetoxyscirpenol **Type-B Trichothecenes** Nivalenol, deoxynivalenol, 3- and 15-acetyldeoxynivalenol, fusarenon-X Moniliformin
	F. graminearum *F. culmorum* *F. sporotrichioides*	Zearalenone
Penicillium	*P. verrucosum* *P. viridicatum*	Ochratoxin (Ochratoxin A)
	P. citrinum *P. verrucosum*	Citrinin
	P. roqueforti	Roquefortine PR toxin
	P. cyclopium *P. camemberti*	Cyclopiazonic acid
	P. expansum *P. claviforme* *P. roquefortii*	Patulin
Neotyphodium (formerly *Acremonium*)	*N. coenophialum*	**Tall fescue toxins:** Ergot alkaloids, lolines, peramine
	N. lolii	**Perennial ryegrass toxins:** Lolitrems, peramine, ergot alkaloid (ergovaline)

1.3 Conditions for fungal growth and mycotoxin production

Mycotoxins occur worldwide. However, geographic and climatic factors affect the production and thus the occurrence of individual mycotoxins (Kuiper-Goodman, 2004). Preferences for a certain temperature range and water activity for growth and mycotoxin production have been observed for some fungal species (Table 1.2) (CAST, 2003; FAO, 2004; Hussein and Brassel, 2001; Marth, 1992; Ribeiro *et al.*, 2006;

Sanchis, 2004; Sweeney and Dobson, 1998). Water activity describes the water availability more precisely than the moisture content and is "the ratio of the water vapor pressure above the grains to that above pure water at the same temperature and pressure" (Mannaa and Kim, 2017: 245).

For instance, for the growth of *F. graminearum*, the optimum temperature and the minimum water activity have been estimated to be 24–26°C and 0.90, respectively (Sweeney and Dobson, 1998). The production of trichothecenes and ZEN by this fungal species is ubiquitous, but more prevalent in warm and moderate climates.

Aspergillus ochraceus grows at temperatures of 8–38°C and produces OTA within the temperature range 12–37°C. Temperature optima for growth and OTA production are 24–37°C and 31°C, respectively.

Penicillium verrucosum grows within the temperature range 0–31°C (optimum 20°C) and at a minimum water activity of 0.80. The temperature optimum for growth and OTA production is 20°C. However, OTA production occurs over the whole temperature range and significant quantities of the toxin can be produced at a temperature as low as 4°C and a water activity as low as 0.86 (Sweeney and Dobson, 1998).

Fusarium verticillioides and *F. proliferatum* growing on corn showed an optimum temperature for FUM production between 15 and 25°C (Samapundo *et al.*, 2005).

Favorable environmental conditions, such as warm temperatures, high rainfall and humidity and a high soil fertility, increase the abundance of *Claviceps* spp. and the production of Ergots (Strickland *et al.*, 2011). Toxic alkaloid-containing ergot sclerotia are used for sexual reproduction and as a resting structure to enable survival under unfavorable conditions (e.g. in temperate zones where they overwinter in the ground and then sexually fruit the following spring when grass hosts are flowering) (CAST, 2003; Kren and Cvak, 1999). Fungal growth continues until the fungus produces this latent structure. Montes-Belmont *et al.*, (2002) reported a mean day temperature of 25°C and a maximum relative humidity of 96% as optimum climatic conditions for ergot development. *C. africana* sclerotia were able to germinate

Table 1.2 – *Preferred temperatures and water activity values for fungal growth and mycotoxin production*

Fungus species	Temperature range for fungal growth (°C)		
	Minimum	Optimum	Maximum
Aspergillus flavus	10–12	25–35	42–43
A. parasiticus	10–12	32–35	42–43
A. ochraceus	8	24–37	37
Penicillium verrucosum	0	20	31–35
Fusarium verticillioides	2–5	23–30	32–37
F. proliferatum	4	30	37
F. culmorum	0–10	20–25	31–35
F. poae	5–10	20–25	35
F. avenacum	5–10	20–25	35
F. tricinctum	5–10	20–25	35
F. graminearum	–	24–26	–
F. sporotrichioides	-2	21–28	35
Claviceps purpurea	9–10	18–22	–

Table 1.2 – *Contd.*

Fungus species	Temperature range for mycotoxin formation (°C)		
	Minimum	Optimum	Maximum
Aspergillus flavus	12–15	30–33	37–40
A. parasiticus	12	33	40
A. ochraceus	12–15	25–31	37
Penicillium verrucosum	4	20–25	–
Fusarium verticillioides	10	15–30	37
F. proliferatum	10	15–30	37
F. culmorum	11	29–30	–
F. graminearum	11	29–30	–
Claviceps purpurea	–	18–20	–

Fungus species	Water activity (aw) for fungal growth		
	Minimum	Optimum	Maximum
Aspergillus flavus	0.80	0.95–0.99	–
A. parasiticus	0.83–0.84	0.95–0.99	–
A. ochraceus	0.77–0.79	0.95–0.99	–
Penicillium verrucosum	0.80	0.95	–
Fusarium verticillioides	0.87–0.90	–	0.99
F. proliferatum	0.90	–	–
F. culmorum	0.90–0.91	0.98–0.99	–
F. poae	0.90–0.91	0.98–0.99	–
F. avenacum	0.90–0.91	0.98–0.99	–
F. tricinctum	0.90–0.91	0.98–0.99	–
F. graminearum	0.90	–	0.99
F. sporotrichioides	0.88	–	0.99

Fungus species	Water activity (aw) for mycotoxin formation		
	Minimum	Optimum	Maximum
Aspergillus flavus	0.82	0.99–0.99	0.99
A. parasiticus	0.87	0.99	–
A. ochraceus	0.80–0.85	0.98	–
Penicillium verrucosum	0.83–0.86	0.90–0.95	–
Fusarium verticillioides	0.92–0.93	–	–
F. proliferatum	0.93	–	–
F. graminearum	0.90–0.91	0.98	–

after 1 year of dry storage at ambient temperature (15–30°C) (Frederickson *et al.*, 1991). *Claviceps africana* colonizes sorghum in Southern Africa, Southeast Asia, South America and the USA (Kren and Cvak, 1999). *Claviceps purpurea* occurs in every temperate region and shows the widest host range of all *Claviceps* spp. (Kren and Cvak, 1999).

1.4 Chemical stability of mycotoxins

Owing to their chemical structure and low molecular weight mycotoxins are chemically stable. They resist high temperatures and various manufacturing processes (Bullerman and Bianchini, 2007). The fate of mycotoxins during thermal food processing has been reviewed (Kabak, 2009).

Aflatoxins have a melting point of between 268 and 269°C and show a high resistance to dry heat-up temperatures. Temperatures above 150°C are required to attain partial destruction of these toxins (Samarajeewa *et al.*, 1990).

Ochratoxin A has a melting point of 169°C. While dry heating of wheat at 100°C for 40–160 min had no effect on OTA, wet heating at the same temperature for 120 min resulted in the destruction of 50% of the toxin (Boudra *et al.*, 1995).

Fumonisins resist temperatures up to 100–120°C (Humpf and Voss, 2004) and may therefore withstand most of the commonly applied thermal processes. It is also known that FUM can bind to various components of the feed matrix or react with other ingredients of the feed. The formation of unidentified biologically active decomposition products may lead to an underestimation of the feed's toxic potential (see also Chapter 1.5).

Deoxynivalenol is a heat resistant compound with a melting point of 151–153°C. Thermal processing did not lead to a significant reduction of DON levels (CAC, 2003b).

Zearalenone is stable during storage, milling and cooking and has a melting point of 164–165°C (EFSA, 2004d). It was shown to withstand exposure to 140°C for 4 h (Smith *et al.*, 1994), but complete destruction was observed within less than 30 min when ZEN was incubated in aqueous buffer solutions at 225°C.

Ergot alkaloids were relatively stable during processing of flour into pasta and oriental noodles (Fajardo *et al.*, 1995). During baking of a rye roll the ergot content decreased by approximately 25% (Bürk *et al.*, 2006).

Overall, it should be kept in mind that in most studies investigating the stability of mycotoxins only the disappearance of the mycotoxins is evaluated. This does not necessarily mean that the toxicity is reduced. Decomposition or transformation products may be just as dangerous as the parent molecules.

1.5 Masked mycotoxins

In defense against their toxic effects, some plants alter the chemical structure of mycotoxins by attaching a hydrophilic residue such as glucose (reviewed by Berthiller *et al.*, 2013). After ingestion of a plant-derived product, the mycotoxin conjugate may be transformed back into the parent mycotoxin in the mammalian digestive tract, as exemplified for deoxynivalenol-3-glucoside (D3G) in Figure 1.1. The conjugated forms are therefore potentially as hazardous to the consumer as their parent compounds. Plant-derived mycotoxin conjugates escape the detection by routinely used analytical methods. The aim of these methods is the detection of the respective parent mycotoxins. The plant-derived conjugates were

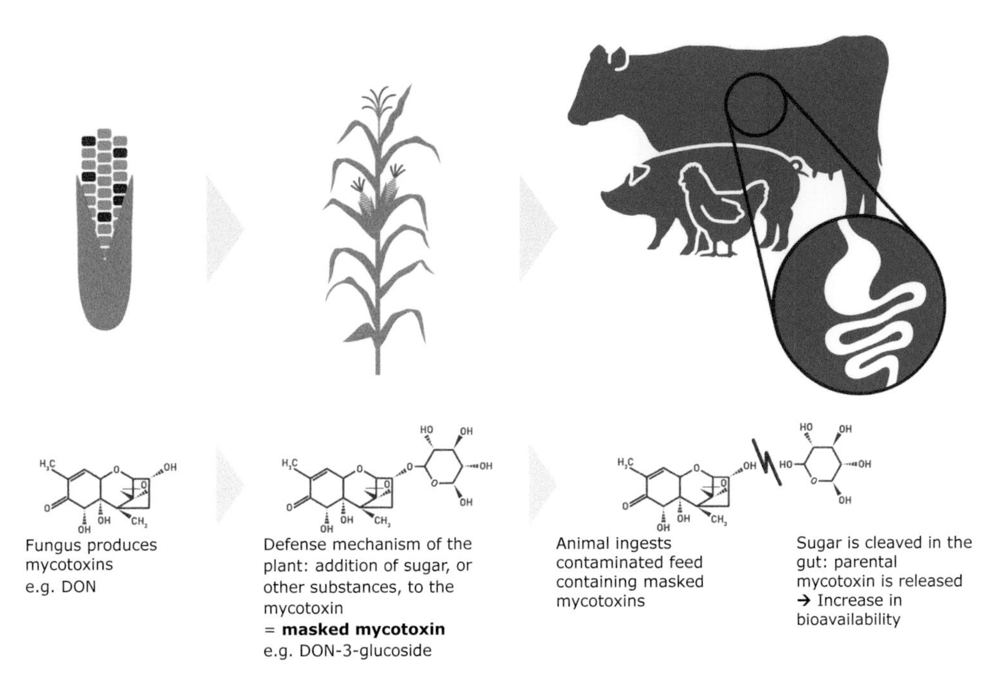

***Figure 1.1** – Scheme of mycotoxin conjugate formation in plants and mycotoxin release in the mammalian digestive tract.*

Note: DON = deoxynivalenol.

therefore designated "masked mycotoxins". These substances exhibit unexpected different physicochemical behavior than their parent compounds and their extraction might be limited when less polar solvents are used (as used for unmodified mycotoxins) (Berthiller *et al.*, 2013). Studies investigating the amount of certain masked mycotoxins in commodities have shown that they may occur frequently and reach considerable levels (e.g. De Boevre *et al.*, 2012; Khaneghah *et al.*, 2018; Streit *et al.*, 2013).

The best-studied example of a masked mycotoxin is the plant-derived mycotoxin conjugate D3G. DON-3-glucoside shows a dramatically reduced toxicity compared with DON (Pierron *et al.*, 2016a; Poppenberger *et al.*, 2003). However, different strains of gut bacteria were shown to hydrolyze D3G to DON *in vitro*, suggesting that DON may be formed in the gut of animals that consumed D3G contaminated feed (Berthiller *et al.*, 2011). Indeed, dietary D3G was shown to be converted to DON in the digestive tract of pigs (Broekaert *et al.*, 2016; Nagl *et al.*, 2014). A fraction of the dietary D3G was subsequently absorbed as DON. Consequently, D3G adds to the DON load of pig diets. In an extensive survey, D3G was detected in 75% of analyzed samples of feed and feed components (Streit *et al.*, 2013). DON-3-glucoside was furthermore shown to reach levels similar to those of DON in cereals (Berthiller deoxynivalenol *et al.*, 2013). In 2016, 359 wheat samples from China were analyzed. Deoxynivalenol was the most prevalent mycotoxin and found in 97% of the samples, but D3G was also found in 33% of all samples (Khaneghah *et al.*, 2018). It was suggested to routinely monitor D3G in cereals, as *Fusarium* resistance breeding might even enhance its incidence in the future (Berthiller *et al.*, 2011).

Besides masked mycotoxins, there are also other categories of mycotoxins, which are not readily detectable by standard analytical methods but potentially toxic to consumers (Rychlik *et al.*, 2014). These include mycotoxin conjugates formed by fungi, by animals or during food processing, as well as mycotoxins that are covalently or non-covalently bound to carbohydrate or protein matrices in food and feed (reviewed by Berthiller *et al.*, 2009, 2013).

1.6 Emerging mycotoxins

One definition of "emerging mycotoxins" describes them as mycotoxins that are increasingly found but until now not routinely determined and not regulated by legislation (Gruber-Dorninger *et al.*, 2017). Some substances listed in Table 1.1 are so called "emerging mycotoxins", for example moniliformin. Detection of many more fungal metabolites of unknown toxicity is now possible due to the availability of advanced analytical methods.

The BIOMIN Mycotoxin Survey analyses mycotoxin occurrence in feed and feed raw materials worldwide. In 2017, 91% of the corn samples analyzed for this survey ($n = 905$) were contaminated with moniliformin. These samples were analyzed using the Spectrum 380® method (see Chapter 5.3.3.). Just recently, EFSA (2018) has published its scientific opinion on health risks related to moniliformin. Toxic effects were observed in animals including rats, pigs, poultry and fish and moniliformin was shown to mainly impair the cardiovascular system. *In vitro* studies showed a clastogenic effect leading to chromosomal damage. However, moniliformin concentrations in animal feed are insufficient to reach doses that caused toxic effects in these studies. Thus, EFSA concluded that there is a low risk for farm animals and also for human health. Still, the report encourages more studies and highlights that especially long-term studies are needed. Additionally, *in vivo* studies on the genotoxicity of moniliformin should be performed and effects on not yet tested animals (e.g. ruminants) should be determined. The emerging mycotoxins beauvericin and enniatins showed toxicity in *in vitro* tests, but only limited toxicity in *in vivo* tests (Gruber-Dornigner *et al.*, 2017). For fusaproliferin, toxic effects have been reported on chicken embryos and shrimp larvae. Fusaric acid showed neurochemical effects and might act synergistically with DON.

1.7 Mycotoxin co-contamination

Feedstuffs are usually contaminated with more than one mycotoxin. There may be many reasons for mycotoxin co-contamination. Plants are commonly contaminated with more than one fungus and many fungal species produce multiple mycotoxins. The blending of different commodities for the creation of a complete diet further increases the likelihood of mycotoxin co-contamination. If blended commodities originate from different geographical regions that are home to different fungal species – a common situation given the global trading of commodities – mycotoxin co-contamination of feedstuffs becomes even more likely.

Comprehensive surveys of mycotoxin content in feedstuffs indicate that mycotoxin co-contamination is the rule rather than the exception. Rodrigues and Naehrer (2012) analyzed 7049 feed samples collected from different geographical regions for the presence of major mycotoxins (Afla, ZEN, FUM and OTA). They found that 48% of the samples contained two or more mycotoxins. Streit *et al.*, (2013) investigated the mycotoxin content of 83 feed samples of diverse origin using a state-of-the-art liquid chromatography-mass spectrometry (LC-MS)/mass spectrometry (MS) method for the parallel detection of 320 fungal metabolites. They found that all analyzed samples were co-contaminated with at least seven mycotoxins or other potentially toxic fungal metabolites. In the BIOMIN Mycotoxin Survey in 2017, 71% of all samples, which were tested for at least three mycotoxins (n = 13,363), were co-contaminated and contained two or more mycotoxins.

1.8 Mycotoxin interactions

Since plant proteins are increasingly being used in feed for aquaculture, the awareness of mycotoxin contamination has grown (Gonçalves *et al.*, 2017). For a thorough evaluation of feed quality, not only contamination levels of individual mycotoxins, but also the co-occurrence of mycotoxins should be considered. Although each individual mycotoxin may only be present at a low level, interactions of mycotoxins can enhance the toxicological effect of the diet.

Different categories of interactions occur, namely additive, less than additive, synergistic and antagonistic interactions. An interaction is additive if the effect of the mycotoxin mixture equals the sum of the effects of the individual toxins, whereas less than additive means only effects of one of the mycotoxins are visible. Synergistic interactions occur if the effect of the mycotoxin mixture is greater than expected based on the effects of the individual toxins. This includes cases in which only the effect of one of the mycotoxins is visible, but much stronger than expected. It describes also cases in which "new" effects, normally not provoked by the individual mycotoxins, are observed. Antagonistic effect means that the effect of at least one of the toxins is reduced (Grenier and Oswald, 2011).

Grenier and Oswald (2011) reviewed publications that investigated the effects of mycotoxin co-contamination *in vivo* in different animal species. They found that all types of mycotoxin interactions – additive, less than additive, synergistic and antagonistic – were frequently encountered. Thus, a combination of different mycotoxins might negatively affect animals even if the concentrations of the individual mycotoxins do not reach levels considered as detrimental (Grenier and Oswald, 2011). The toxicity of mycotoxins can be investigated by analyzing different parameters, for instance animal performance, various biochemical parameters, response to pathogens, histopathology and so forth. The type of interaction may be different for each parameter. It may be additive for one parameter, but synergistic for another. The type of interaction can be due to many factors such as nutritional state of the animal, age, sex and more. Importantly, Grenier and Oswald (2011) concluded that co-exposure to multiple toxins finally caused greater total effects than exposure to each individual toxin, even in cases where less than additive or antagonistic effects were detected. Consequently, co-contamination of feed with mycotoxins does always result in a higher risk to animals.

Mycotoxin interactions are especially relevant in the case of fusariotoxins because they often co-occur. *Fusarium graminearum* and *F. culmorum* are known to produce several different toxins under the same conditions, for example, ZEN and DON, which have been shown to interact synergistically. Deoxynivalenol is furthermore often found to co-occur with other trichothecenes (T-2, nivalenol [NIV], diacetoxyscirpenol [DAS]) and with FUM. Only few studies have explored synergistic or additive effects of mycotoxins on aquaculture species and those were all investigating fish. The studies and results will be described in Chapter 3. A short overview can be seen in Table 1.3. Figure 1.2 also illustrates the current knowledge on the interactions of major mycotoxins in fish.

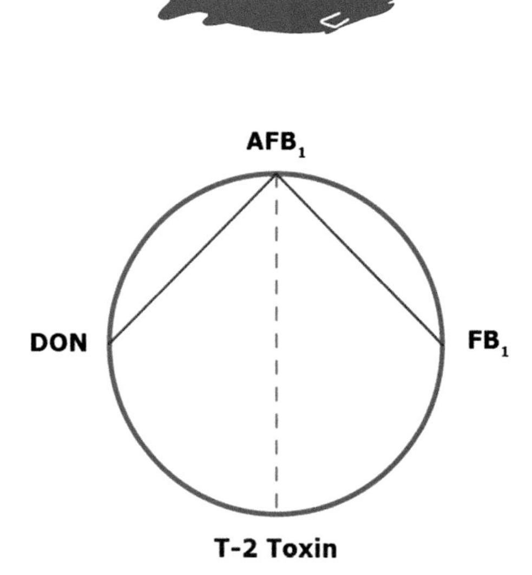

Figure 1.2 *– Synergistic (solid line) and additive (dashed line) effects of mycotoxins in fish.*

Note: Afla B_1 = aflatoxin B_1; DON = deoxynivalenol; FB_1 = fumonisin B_1.

Table 1.3 *– Mycotoxin combinations in fish*

Mycotoxins	Species tested	Effect	References
Afla B_1 + FB_1	Trout	Synergistic	Carlson *et al.* (2001)
Afla B_1 + T-2 toxin	Mosquitofish	Additive	McKean *et al.* (2006a)
DON + Afla B_1	Carp	Synergistic	He *et al.* (2010)

Note: Afla B_1 = aflatoxin B_1; DON = deoxynivalenol; FB_1 = fumonisin B_1.

1.9 Mode of action/toxicology/metabolism of mycotoxins

As their chemical structures vary considerably, mycotoxins cannot be classified as one group according to their mode of action, toxicology or metabolism. In the following pages, the most important mycotoxins and their modes of action, toxicology and metabolism as well as main symptoms and target organs will be described. It is important to bear in mind that additive and synergistic effects can occur in the presence of two or more mycotoxins.

1.9.1 Aflatoxins

1.9.1.1 General aspects

Aflatoxins were identified in 1960 and represent one of the most thoroughly studied group of mycotoxins. They are relatively hydrophilic and mainly produced by certain strains of *Aspergillus parasiticus* and

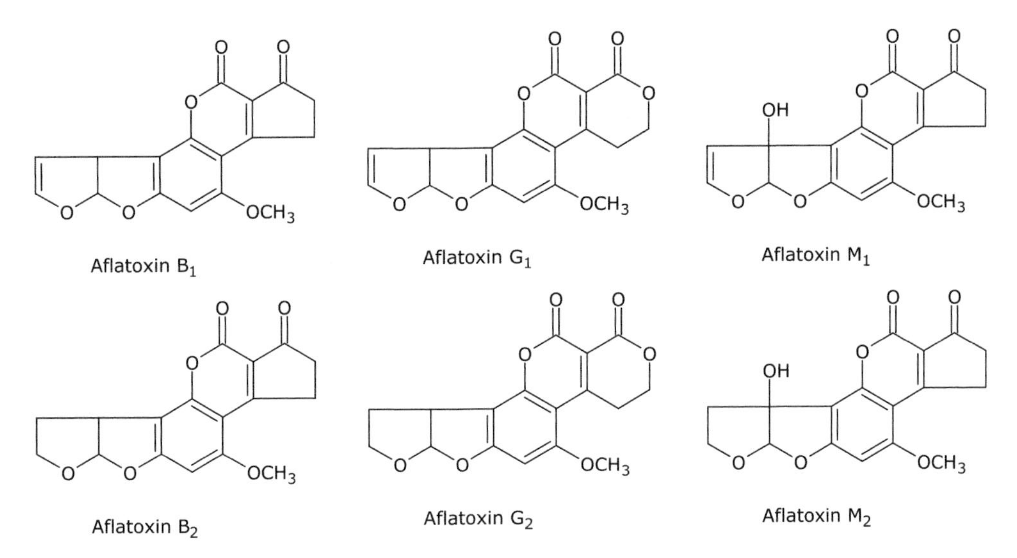

***Figure 1.3** – Chemical structures of the major aflatoxins.*

Aspergillus flavus, and mainly occur in agricultural products originating from tropical or subtropical regions (Vila-Donat *et al.*, 2018). There are six major Afla (Figure 1.3) that were named according to their fluorescent properties under ultraviolet light (c. 365 nm) and their chromatographic mobility (subscripts). Aflatoxins B_1 (Afla B_1) and B_2 (Afla B_2) fluoresce blue while aflatoxins G_1 (Afla G_1) and G_2 (Afla G_2) fluoresce green. The metabolic products of Afla B_1 and Afla B_2 – aflatoxin M_1 (Afla M_1) and M_2 (Afla M_2) – occur in the milk of lactating mammals after the consumption of aflatoxin contaminated feed. Aflatoxin B_1 is the most toxic and most prevalent species of Afla.

1.9.1.2 Exposure and absorption into the organism

Because of the common occurrence of Afla in feedstuffs, feeds and milk products, these mycotoxins pose a serious threat to humans and animals. Ingestion is the main route of exposure, but inhalation is relevant as well if people or animals are exposed to grain dust. After respiratory exposure, Afla B_1 may appear in the blood more quickly than after oral exposure. After 4 h, the plasmatic concentration does not differ between the two routes of exposure. Following ingestion, Afla B_1 is efficiently absorbed from the gastrointestinal tract. The duodenum appears to be the major site of absorption. Owing to the compound's low molecular weight, the main mechanism of absorption is passive diffusion. Once Afla B_1 has entered the blood stream, it is transported to the liver, the major site of metabolism (Gratz, 2007).

1.9.1.3 Metabolism

The metabolism of Afla B_1 has been extensively reviewed (Eaton *et al.*, 2010; International Agency for Research on Cancer [IARC], 1993, 2002), and can be divided into three phases:

- I bioactivation
- II conjugation
- III deconjugation.

I Bioactivation

Phase I bioactivation is a prerequisite for Afla B_1 toxicity. In the liver and in other tissues Afla B_1 is metabolized by cytochrome P450 enzymes to aflatoxin P_1 (Afla P_1), Afla M_1, aflatoxin Q_1 (Afla Q_1) and Afla B_1–8,9-epoxide (Figure 1.4) (Riley and Voss, 2011). While Afla Q_1 shows a low toxicity, Afla M_1 is acutely toxic and Afla B_1–8,9-epoxide is acutely toxic, mutagenic and carcinogenic. Afla B_1–8,9-epoxide is highly unstable and therefore reacts with the following molecules in its environment (Eaton *et al.*, 2010).

- Biological nucleophils (such as nucleic acids): Afla B_1–8,9-epoxide forms stable links to DNA thereby inducing point mutations and strand breaks. The formation of Afla B_1-DNA adducts is highly correlated with the carcinogenic effect of Afla B_1 in animals and humans.
- Water: In the presence of water molecules Afla B_1–8,9-epoxide is hydrolyzed to Afla B_1–8,9-dihydrodiol, which forms adducts with serum proteins, such as albumin. Albumin is the most abundant plasma protein and synthesized in the liver. It is important for osmotic pressure in the blood and thus water exchange, as well as for the transport of many substances including organic anions, hormones, etc. This mechanism may explain some of the toxic effects of aflatoxin.

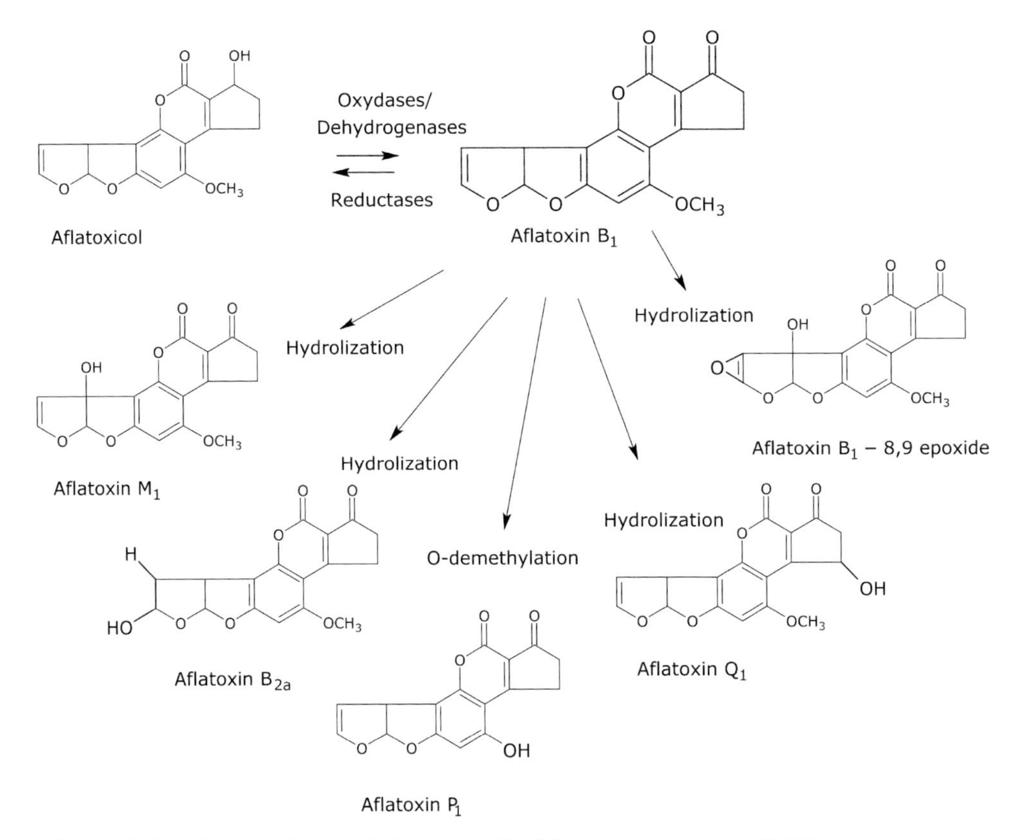

***Figure 1.4** – Aflatoxin B_1 metabolism (modified from Leeson* et al., *1995).*

II Conjugation

Phase I metabolites undergo phase II biotransformation, which involves the formation of glutathione, glucuronide and sulfate conjugates by the action of the enzymes glutathione S-transferase (GST), ß-glucuronidase and sulfate transferase, respectively. The major identified conjugate of Afla B_1–8,9-epoxide is the glutathione conjugate. This conjugation is the principal detoxification pathway of activated Afla B_1 in many mammals and it is essential in the reduction and prevention of Afla B_1 induced carcinogenicity. The resulting conjugates are excreted via the bile into the intestinal tract. GST activity is inversely correlated with the susceptibility of several animal species to Afla B_1 induced carcinogenicity (Figure 1.5).

III Deconjugation

Bacteria in the intestinal tract may catalyze deconjugation of aflatoxin conjugates. Deconjugation may result in reabsorption of toxic compounds and in the establishment of an enterohepatic circulation.

1.9.1.4 Excretion and residues in animal products

Aflatoxin B_1 and its metabolites are mainly excreted via bile and urine. In lactating animals, Afla M_1 and other metabolites are excreted in the milk. Many studies showed the carry-over of Afla into animal products, such as porcine tissue, milk and milk products (Giovati *et al.*, 2015; Völkel *et al.*, 2011). A recent study analyzing milk products in Kenya found a high incidence of Afla M_1. Although 50% of the samples

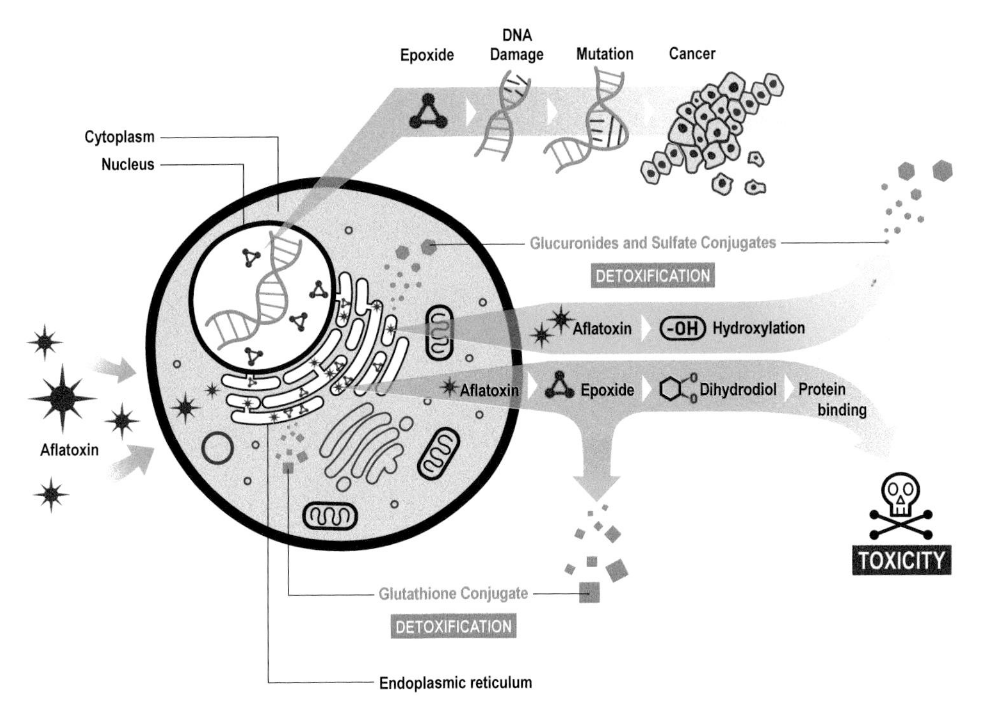

Figure 1.5 *– Simplified mode of action of aflatoxin B_1 (adapted from Riley and Voss, 2011).*

showed higher concentrations than allowed in the European Union (EU), only three samples exceeded the limits set in the USA (Lindahl *et al.*, 2018). In studies from different EU countries, the abundance of Afla M_1 concentrations in milk were low (only 0.06% exceeded the European maximum concentrations) (Marin *et al.*, 2013).

Besides Afla M_1, Afla B_1-DNA adducts and Afla B_1-albumin adducts are currently available biomarkers for Afla B_1 exposure (Baldwin *et al.*, 2011).

1.9.1.5 Toxicity

Aflatoxins were classified as group 1 carcinogens (carcinogenic to humans) by the IARC (2012).

In animals, the effects of Afla vary according to species, sex, age and even animal breed. Variations in the expression level of the glutathione transferase system and alterations in the cytochrome P450 system are thought to contribute to the differences observed in the susceptibility of animals to Afla (Bennett and Klich, 2003; Eaton *et al.*, 2010).

For humans and animals, the main target organ of Afla is the liver. Nevertheless, several organs like kidneys, reproductive and immune systems are targeted. The main symptoms include vomiting, necrosis, anorexia, fatty liver, liver cancer and diarrhea. In humans, several effects on the reproductive system have been reported: morphological changes and delayed development of testicles, a decrease in the percentage of live sperm and reduced plasma concentration of testosterone (Marin *et al.*, 2013). Furthermore, a correlation between exposure to Afla M_1 present in milk and reduced growth of children has been suggested by studies in Kenya and Iran (Lindahl *et al.*, 2018). Stunted development of children was also associated with aflatoxin exposure in other studies, but the mechanisms are not understood yet (Logrieco *et al.*, 2018).

The effects on domestic animals are not limited to toxic hepatitis and jaundice but involve a broad range of organs, tissues and systems (Table 1.4). Immunosuppressive effects comprise less resistance to secondary infections by fungi, bacteria and parasites. A decrease in weight gain and an increase in the feed conversion ratio (FCR) are frequently observed. Exposure can also lead to death of the affected animals.

1.9.2 Trichothecenes

1.9.2.1 General aspects

Trichothecenes are a family of more than 200 structurally related compounds. The structure of trichothecenes is characterized by a sesquiterpene ring and a C-12,13-epoxide ring (Figures 1.6 and 1.7). Trichothecenes are mainly produced by several *Fusarium* spp. (Pestka, 2010a). Other fungal species, such as *Stachybotrys* and *Myrothecium*, also produce certain trichothecenes. Most trichothecenes produced by *Fusarium* spp. are categorized as type-A or type-B trichothecenes:

- **type-A trichothecenes** (namely T-2 toxin, HT-2 toxin [HT-2], DAS) contain a functional group other than a ketone at the C-8 position (Figure 1.6 and Table 1.5)
- **type-B trichothecenes** (namely DON, 3- and 15-acetyldeoxynivalenol, NIV, fusarenon X) contain a carbonyl function at the C-8 position (Figure 1.7 and Table 1.6).

Deoxynivalenol, also known as vomitoxin, is the most frequently occurring trichothecene. T-2 toxin and DAS are soluble in non-polar solvents, whereas DON and NIV are soluble in polar solvents like alcohol and water.

***Table 1.4** – Main systems affected by aflatoxins. (All animal species are included in this table. For the effects of aflatoxins in aquaculture, please consult Chapter 3.)*

Affected system	Effects/signs/symptoms
Genes/gene expression	Teratogenic effects – birth defects
	Carcinogenic effects – higher incidence of cancer in exposed animals
Hepatotoxic effects	Weight variation of the liver, fatty liver syndrome, change in the texture and coloration, jaundice
Circulatory system	Hematopoietic effects (hemorrhages, anemia)
Nephrotoxic effects	Enlargement of the kidneys
Immune system	Immunosuppression (decreased resistance to environmental and microbial stressors, increased susceptibility to diseases), Changes in the bursa of Fabricius and thymus reduction, Weight variation of the spleen
Nervous system	Nervous syndrome (e.g. abnormal behavior)
Skin	Dermatoxic effects (impaired feathering)
Urinary system	Kidney inflammation
Digestive system	Impaired rumen function with decreased cellulose digestion, decreased volatile fatty acid formation, decreased proteolysis, decreased rumen motility, diarrhea; Change in the texture and coloration of the gizzard
Reproductive system	Decreased breeding efficiency (birth of smaller and unhealthy offspring)

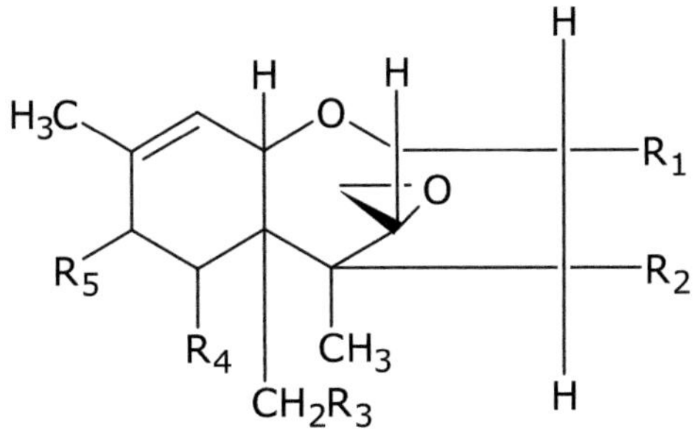

***Figure 1.6** – Structural formula of type-A trichothecenes.*

1.9.2.2 Mechanism of action

Trichothecenes are potent inhibitors of protein biosynthesis. Their 12,13-epoxide ring structure and other functional groups bind to the 60S ribosomal subunit and interact with the peptidyltransferase (Pierron *et al.*, 2016b). They thereby impair the initiation (e.g. T-2, HT-2, DAS) or elongation (e.g. DON)

***Table 1.5** – Structural formula of type-A trichothecenes*

	Molecular formula	R1	R2	R3	R4	R5
Diacetoxyscirpenol	$C_{19}H_{26}O_7$	OH	OAc	OAc	H	H
T-2 toxin	$C_{24}H_{34}O_9$	OH	OAc	OAc	H	$OCOCH_2CH(CH_3)_2$
HT-2 toxin	$C_{22}H_{32}O_8$	OH	OH	OAc	H	$OCOCH_2CH(CH_3)_2$

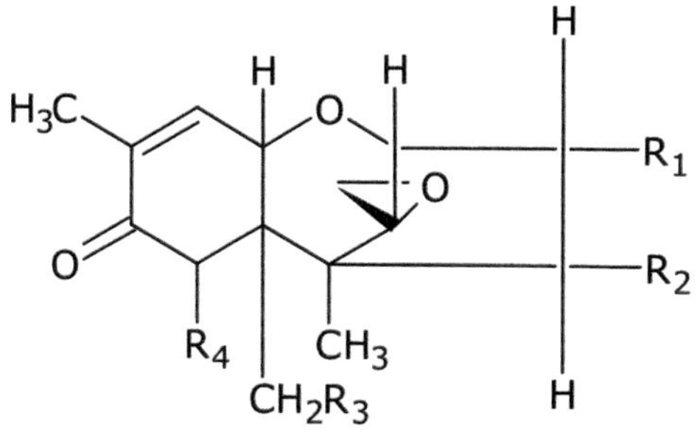

***Figure 1.7** – Structural formula of type-B trichothecenes.*

***Table 1.6** – Structural formula of type-B trichothecenes*

	Molecular formula	R1	R2	R3	R4
Deoxynivalenol	$C_{15}H_{20}O_6$	OH	H	OH	OH
3-Acetyldeoxynivalenol	$C_{17}H_{22}O_6$	OAc	H	OH	OH
15-Acetyldeoxynivalenol	$C_{17}H_{22}O_6$	OH	H	OAc	OH
Nivalenol	$C_{15}H_{20}O_7$	OH	OH	OH	OH
Fusarenon X	$C_{17}H_{22}O_8$	OH	OAc	OH	OH

phase of translation. The disruption of protein biosynthesis is followed by a secondary disruption of RNA and DNA synthesis.

Ribosome damage by DON induces a ribotoxic stress response (Pestka *et al.*, 2004, 2007, 2010b). The ribosome-associated proteins RNA-activated protein kinase (PKR) and hematopoietic cell kinase (HCK) are activated. The subsequent phosphorylation of mitogen-activated protein kinases (MAPKs) drives the activation of transcription factor (TF), induction of inflammatory cytokine expression and apoptosis and eventually leads to chronic and immunotoxic effects (Figure 1.8). Trichothecenes are very cytotoxic to eukaryotic cells, causing cell lysis and inhibition of mitosis. They are especially toxic to tissues with a high cell division rate such as the intestinal mucosa and lymphoid tissue.

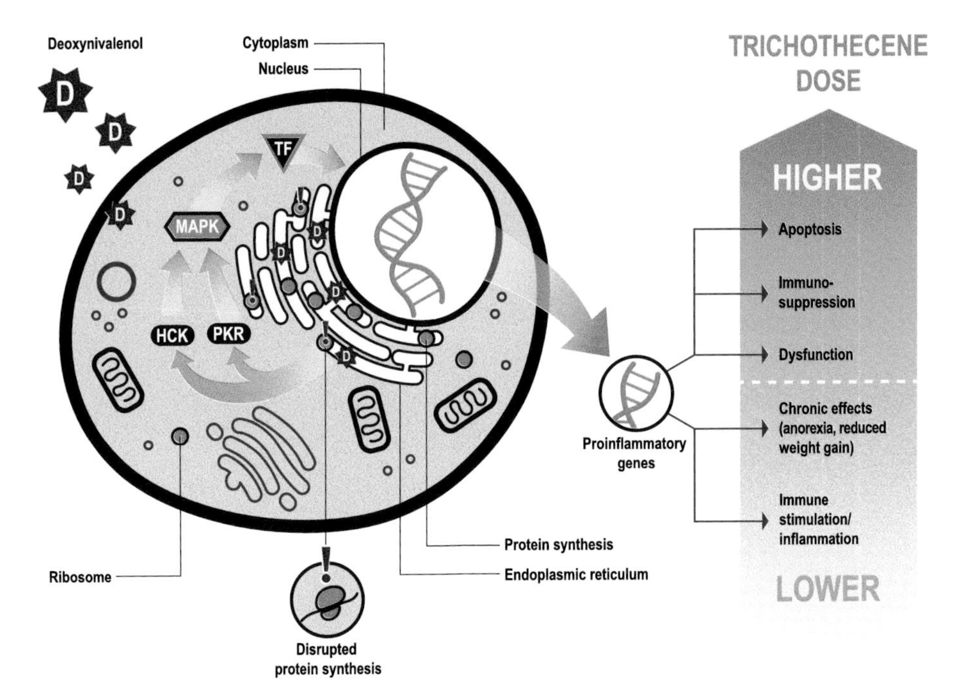

***Figure 1.8** – Mechanism of action of deoxynivalenol (D) (modified from Pestka* et al., *2004 and Pestka, 2007).*

Note: HCK – hematopoietic cell kinase, PKR – RNA-activated protein kinase, MAPKs – mitogen-activated protein kinases, TFs – transcription factors.

1.9.2.3 Metabolism

Generally, there are three main metabolic pathways:

- I conjugation
- II de-epoxidation
- III de-acetylation.

The de-epoxidation is the most important step in the detoxification of trichothecenes and may be carried out by certain microorganisms in the gastrointestinal tract of ruminants.

1.9.2.4 Absorption/residues

In general, DON is rapidly absorbed from the digestive tract and widely distributed in many tissues and organs. Deoxynivalenol residues have until now not been found in tissues of aquaculture species investigated, but in pig tissues in different studies, albeit at very low levels (Döll *et al.*, 2008; Goyarts *et al.*, 2007). Mycotoxin residues in animal tissues are very important as they pose a risk to human consumers.

1.9.2.5 Toxicity

Several outbreaks of disease in humans including symptoms like vomiting, gastrointestinal disorders, diarrhea or headaches have been attributed to the consumption of *Fusarium*-contaminated grains (Pestka, 2010a). Long-term exposure to trichothecenes can provoke a disease named alimentary toxic

Table 1.7 *– Main systems affected by trichothecenes. (All animal species are described in this table. For the effects of trichothecenes on species in aquaculture, please see Chapter 3.)*

Affected system	Effects/signs/symptoms
Circulatory system	Hematopoietic effects (hemorrhages; blood pattern disorders), necrosis of hematopoietic tissue
Immune system	Immunosuppression (decreased resistance to environmental and microbial stressors; increased susceptibility to diseases), necrosis of the lymphoid tissue
Digestive system	Gastro-intestinal effects: gastroenteritis, swelling of stomach and intestines, inflammation of the rumen; lesions in the intestinal tract; vomiting; feed refusal – anorexia; diarrhea, gizzard lesions
Reproductive system	Decreased breeding efficiency (birth of smaller and unhealthy offspring)
Nervous system	Neurotoxic effects (restlessness; lack of reflexes; abnormal wings positioning; nervous syndrome)
Skin	Dermotoxicity (oral and dermal lesions; necrosis)

aleukia (ATA) in humans (Marin *et al.*, 2013). The first symptoms of this disease are burning sensations, which can lead to severe gastroenteritis, destruction of bone marrow, immunosuppression, hemorrhages and eventually death by asphyxia or lung bleeding (Marin *et al.*, 2013).

In animals, decrease in feed consumption (anorexia) and vomiting are two characteristic effects of trichothecene intoxication (Table 1.7). Several effects on the gastrointestinal tract have been observed, including gastroenteritis (swelling of stomach and intestines). Another primary target of trichothecenes is the immune system. Swine are considered to be the most sensitive livestock species (Logrieco *et al.*, 2018).

Deoxynivalenol is less toxic than other trichothecenes, but as it occurs very frequently in cereals worldwide, it is nevertheless the most relevant trichothecene. Frequently observed symptoms of DON intoxication include feed refusal, reduced weight, vomiting, bloody diarrhea, lesions in the intestinal tract, hemorrhaging and immunosuppression (Vila-Donat *et al.*, 2018).

All effects observed in aquaculture species are discussed in Chapter 3.

1.9.3 Ochratoxins

1.9.3.1 General aspects

Ochratoxins are metabolites of storage fungi *Aspergillus ochraceus* and *Penicillium verrucosum*, occur in temperate regions and are present in a large variety of feeds and foods. There are four types of ochratoxins (A, B, C and D). OTA is the most relevant ochratoxin. It is a contaminant of cereals, beans, rice, coffee, cocoa beans and other plant products (Patriarca and Pinto, 2017). The most significant effect of ochratoxins in farm animals is nephrotoxicity (Pfohl-Leszkowicz and Manderville, 2007).

Chemically, ochratoxins contain an isocoumarin moiety linked to phenylalanine with a peptide bond. They are soluble in ethanol, methanol and acetone and to some degree in water (Malir *et al.*, 2016).

1.9.3.2 Mechanism of action

Ochratoxin A`s nephrotoxity and other toxic effects are likely based on the inhibition of protein synthesis. Ochratoxin A inhibits the enzyme involved in the synthesis of the phenylalanine-tRNA complex, and thus prevents correct protein synthesis. It might also interact with other enzymes that use phenylalanine as a substrate, for example the phenylalanine hydroxylase, which catalyzes the irreversible hydroxylation of phenylalanine to tyrosine. Additionally, OTA promotes lipid peroxidation, which leads to lipid degradation and thus damage of cell membranes (Malir *et al.*, 2016). It modulates the mitogen-activated protein (MAP) kinase cascade, which is very important in expression of proteins and thus cell signaling (Malir *et al.*, 2016). The metabolites hydroxyl quinone ochratoxin (OTHQ) and quinone ochratoxin (OTQ) form DNA adducts and thus cause DNA damage. Furthermore, OTQ and other intermediates promote production of reactive oxygen species (ROS), causing DNA damage and general cytotoxicity.

1.9.3.3 Metabolism

Many metabolites of OTA exist. The metabolites show different toxicity. The major metabolite of ochratoxin is ochratoxin α (OTα), a hydrolysis product without the phenylalanine moiety, which is produced by the gut microflora. Ochratoxin α is considered to be a non-toxic product (Wu *et al.*, 2011). Contrary, the intermediate ochratoxin A open lactone (OP-OTA) is considered as more toxic than its parent compound (Pfohl-Leszkowicz and Manderville, 2007; Wu *et al.*, 2011). Other metabolites are the hydroxylated derivatives OTHQ, ochratoxin B (OTB) and so forth (Figure 1.9). Ochratoxin A is probably metabolized into the quinone-type intermediates by cytochrome P450 enzymes (Malir *et al.*, 2016).

1.9.3.4 Absorption/excretion/residues

Between 40 and 66% of OTA is absorbed from the gastrointestinal tract depending on the species. The small intestine has been shown to be the main site of absorption and maximal absorption occurs in the jejunum (Pfohl-Leszkowicz and Manderville, 2007).

Ochratoxin A binds rapidly to serum albumin and is distributed in the blood mainly in this bound form. Generally, the toxin has a long biological half-life due to its high rate of binding to serum protein although differences between species exist. Therefore, the ochratoxin albumin adduct in serum can be used as a biomarker for ochratoxin exposure (Baldwin *et al.*, 2011).

Ochratoxin A primarily accumulates in the kidneys followed by the liver and muscles but was also detected in whole blood and blood plasma (Battacone *et al.*, 2010). Furthermore, OTA has been detected in milk, including human milk, which could pose a threat to infants (Marin *et al.*, 2013; Patriarca and Pinto, 2017).

1.9.3.5 Toxicity

According to IARC, ochratoxins are classified as possible human carcinogens (Group 2b). Tumors in kidney and liver have been found in experiments with rats. Kidney tumors were hypothesized to be a cause of the Balkan Endemic disease (Massoud *et al.*, 2018). This disease is chronic and leads to renal failure (Massoud *et al.*, 2018).

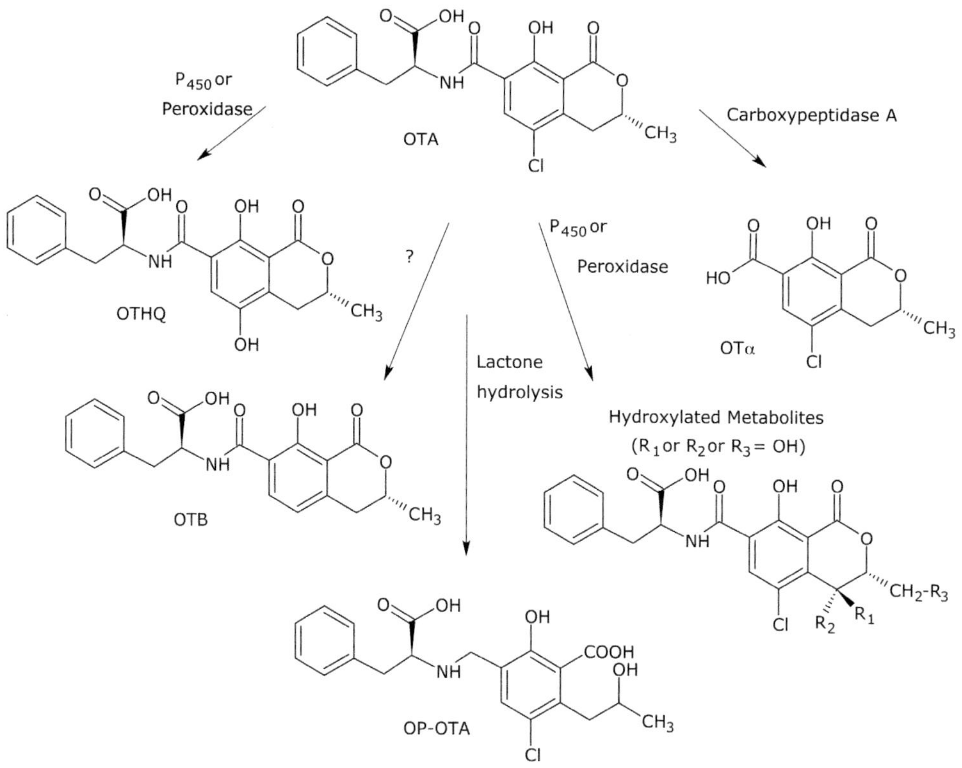

***Figure 1.9** – Metabolism of ochratoxin A (modified from Pfohl-Leszkowicz and Manderville, 2007).*

Note: OTHQ = hydroxyl quinone ochratoxin, OTB = ochratoxin B, OTα = ochratoxin alpha, OP-OTA = ochratoxin A open lactone.

In animals, toxicity varies widely according to animal species and sex. Ochratoxin A has carcinogenic, teratogenic, immunotoxic, embroytoxic, genotoxic, hepatotoxic and possibly neurotoxic effects (Malir *et al.*, 2016; Vila-Donat *et al.*, 2018). The kidneys are the main target organs. Ochratoxin A has been found to cause porcine nephropathy, an extensively studied disease. In the beginning, the nephritic tubule is degenerated and renal interstitial fibrosis occurs. Subsequently, the basal membrane gets thinner and deposits of hyaline, a substance with glassy appearance, occur in the glomeruli (glomerular hyalinization) (Vila-Donat *et al.*, 2018). In rats and mice, OTA has been shown to cross the placenta barrier, harming the embryo (Marin *et al.*, 2013). Immunotoxicity in animals was related to a decrease in size of the thymus, spleen and lymph nodes, less response of macrophages, changes in the number and function of immune cells and modulation of cytokine production (Marin *et al.*, 2013). In pigs, chronic exposure to ochratoxins first leads to reductions of feed consumption and weight gain (Vila-Donat *et al.*, 2018). Ruminants show resistance and are less affected by OTA (Vila-Donat *et al.*, 2018). Furthermore, ochratoxins may also affect other systems as described in Table 1.8.

Table 1.8 – *Main systems affected by ochratoxins. (All animal species are described in this table. For the effects of ochratoxins on aquaculture species, please see Chapter 3.)*

Affected system	Effects/signs/symptoms
Circulatory system	Hematopoietic effects (hematological disorders, blood in urine and feces)
Nephrotoxic effects	Increased water consumption; kidney and urinary bladder dysfunction, renal interstitial fibrosis
Immune system	Immunosuppression (decrease in the size of the thymus, spleen and lymph nodes; less response of macrophages; changes in the number and function of immune cells; modulation of cytokine production; decreased resistance to environmental and microbial stressors; increased susceptibility to diseases)
Hepatotoxic effects	Liver damage
Digestive system	Gastro-intestinal effects (diarrhea)
Reproductive system	Teratogenic, embroytoxic, cross the placenta barrier
DNA damage/genes	Carcinogenic effects

1.9.4 Fumonisins

1.9.4.1 General aspects

Fumonisins are a group of mycotoxins mainly produced by *Fusarium verticillioides* and *F. proliferatum*. These fungi are mostly field fungi and can be found in soil and on plant seeds or residues. They are able to infect intact kernels via their stylar canal and thus the infection might not be recognized (Duncan and Howard, 2010). Damage to kernels, for example by insects, facilitates infection with *Fusarium* spp.

Fumonisins were first isolated in 1988 from culture material of *F. moniliforme* originating from South Africa. They occur worldwide and are frequently detected in corn. This toxin group comprises several analogs including fumonisin B_1 (FB_1), fumonisin B_2 (FB_2) and fumonisin B_3 (FB_3) (Masching *et al.*, 2016). The most prevalent and toxic FUM is FB_1. They are highly polar compounds and soluble in water (Braun and Wink, 2018).

1.9.4.2 Mechanism of action

Fumonisins disrupt the sphingolipid metabolism. Sphingolipids are membrane lipids and important for the structure of cell membranes and lipoproteins as well as for cell regulation (Merrill *et al.*, 2001). Owing to the structural similarity of FUM to the sphenoid bases sphingosine and sphinganine (Figure 1.10), they are competitive inhibitors of ceramide synthase (CerS), a key enzyme in the sphingolipid metabolism. This enzyme catalyzes the acylation of sphinganine (Sa) and sphingosine (So), important reactions in the sphingolipid biosynthesis and salvage pathways, respectively (Merrill *et al.*, 2001). The inhibition of CerS causes an accumulation of Sa, sphinganine-1-phosphate, So and sphingosine-1-phosphate, whereas the level of complex sphingolipids is decreased (Figure 1.11). Free sphenoid bases have a proapoptotic, cytotoxic and growth inhibitory effect on many types of cells (Desai *et al.*, 2002). Many toxic effects

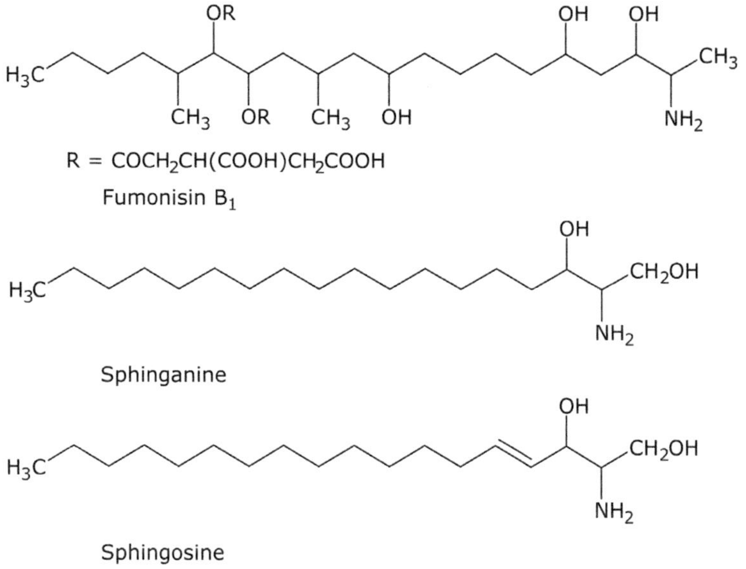

$R = COCH_2CH(COOH)CH_2COOH$

Fumonisin B_1

Sphinganine

Sphingosine

***Figure 1.10** – Structures of fumonisin B_1, sphinganine and sphingosine.*

caused by FUM are due to alterations in the levels of sphingolipids and intermediates of the sphingolipid metabolism in different organs and cell types (Eaton *et al.*, 2010; Riley *et al.*, 1996).

The accumulation of free sphenoid bases in the serum, tissue and urine is a useful biomarker of FUM exposure, because it indicates the extent of disruption of the sphingolipid metabolism (Riley *et al.*, 1993, 1994). Sphinganine and sphinganine-1-phosphate accumulate at a higher rate than So and sphingosine-1-phosphate, respectively. Consequently, the Sa/So and sphinganine-1-phosphate/sphingosine-1-phosphate ratios are the most commonly investigated biomarkers.

1.9.4.3 Exposure and absorption into the organism

Several studies indicate that FUM are poorly absorbed from the gastrointestinal tract and rapidly cleared from the blood. However, the absorbed fraction seems to undergo a wide distribution, with a high affinity to the liver and kidneys, which later on slowly release the toxins (Prelusky *et al.*, 1996).

1.9.4.4 Excretion and residues in animal products

Fumonisins primarily accumulate in kidneys, spleen, liver and lung (Meyer *et al.*, 2003). Fumonisins carry-over to sow milk and pork meat occurs after a high level of exposure over a longer period of time (Völkel *et al.*, 2011).

1.9.4.5 Toxicity

Fumonisin B_1 was classified as a possible human carcinogen (group 2B) by IARC. A connection between the development of esophageal cancer and the exposure to FUM in China and South Africa has been proposed (Marin *et al.*, 2013; Patriarca and Pinto, 2017). Furthermore, neural tube defects have been correlated to FUM exposure around the Mexico–US (Texan) border. A connection between FUM and childhood stunting has also been proposed (Logrieco *et al.*, 2018).

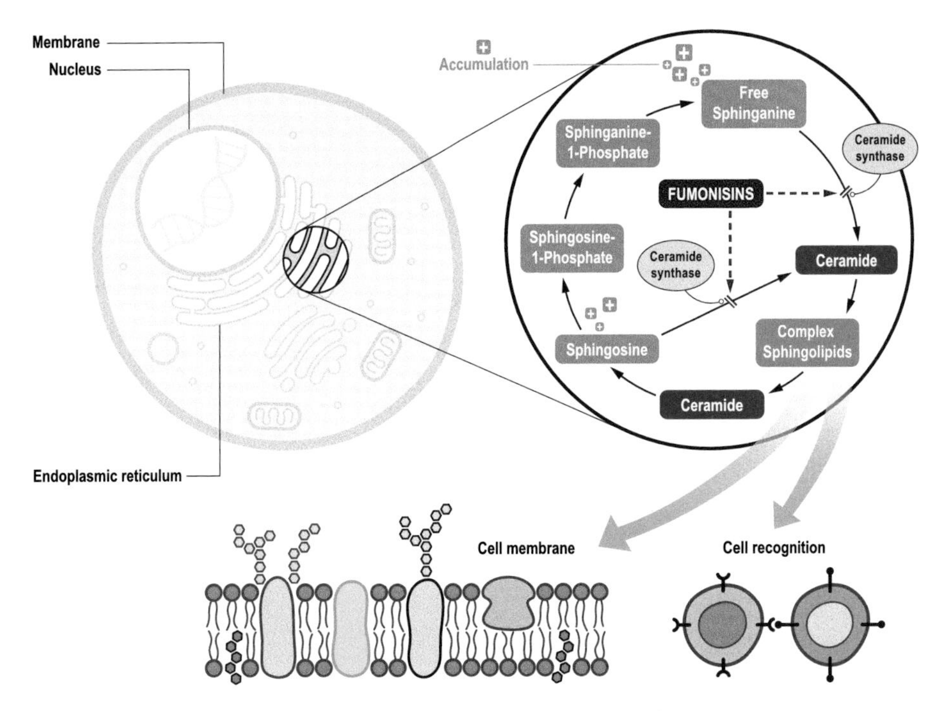

Figure 1.11 *– Sphingolipid metabolism and its disruption by fumonisins. Fumonisins inhibit ceramide synthase. This inhibition causes an increase in the levels of sphinganine and sphingosine, which has a toxic effect on most cells. The Sa/So ratio serves as a biomarker for FUM toxicity. Arrows indicate an increase ↑ or decrease ↓ in the levels of the respective substance due to FUM exposure. For the sake of simplicity, only the main intermediates and enzymes are depicted (modified from Merrill* et al., *2001; Voss* et al., *2007).*

Fumonisins exert toxic effects on the liver, the kidneys and the immune system (Marim *et al.*, 2013; Voss *et al.*, 2007). Furthermore, because of the poor absorption of FUM, the intestines are exposed to the major part of the ingested FUM dose and are therefore a prime target of FUM toxicity (Escriva *et al.*, 2015; Marin *et al.*, 2013; Vila-Donat *et al.*, 2018). Fumonisins cause a broad range of effects in animals (Table 1.9), including cardiotoxicity, dyspnea (shortness of breath), cyanosis (bluish appearance usually caused by low oxgyen levels in the red blood cells), less consumption of feed and weakness (Marin *et al.*, 2013; Vila-Donat *et al.*, 2018). In pigs and horses, FUM cause the species-specific fatal diseases porcine pulmonary edema (PPE) (Haschek *et al.*, 2001) and equine leukoencephalomalacia (ELEM) (Caloni and Cortinovis, 2010; Marin *et al.*, 2013), respectively. Symptoms of PPE are less feed intake, dyspnea, weakness and cyanosis. ELEM, a disease of the central nervous system, leads to lethargy, blindness, less feed consumption, convulsions and very soon to death (Marin *et al.*, 2013; Vila-Donat *et al.*, 2018). Furthermore, FUM exposure was found to compromise gut health (Oswald *et al.*, 2003). A contamination with low or moderate levels of FUM can already negatively affect intestinal health and immune function. A disruption of the intestinal barrier can allow increased translocation of other toxic entities and pathogens (Antonissen *et al.*, 2014).

Table 1.9 *– Main systems affected by fumonisins. (All animal species are described in this table. For the effects of fumonisins on aquaculture species, please see Chapter 3.)*

Affected systems	Effects/signs/symptoms
Immune system	Immunosuppression (decreased resistance to environmental and microbial stressors; increased susceptibility to disease), increase of spleen weight
Digestive system	Intestinal lesions, decreased barrier function of intestinal epithelium (increased translocation of other toxic entities and pathogens), pancreatic necrosis
Circulatory system	Hematopoietic effects (hematological disorders, increased concentration of hemoglobin, cyanosis), cardiotoxic effects
Nervous system	Neurotoxic effects
Hepatotoxic effects	Liver damage, increase of liver weight
Nephrotoxic effects	Kidney damage, increase of kidney weight

1.9.5 Zearalenone

1.9.5.1 General aspects

Zearalenone (Figure 1.12) is an important mycotoxin occurring in warm and temperate climate regions. It is produced mainly by *Fusarium graminearum*, *F. culmorum*, *F. cerealis*, *F. equiseti* and *F. verticillioideson*. *Fusarium* fungi are field fungi and found on a variety of cereal crops. Growth has also been observed under storage conditions. Zearalenone is often found in co-occurrence with the other *Fusarium* toxins

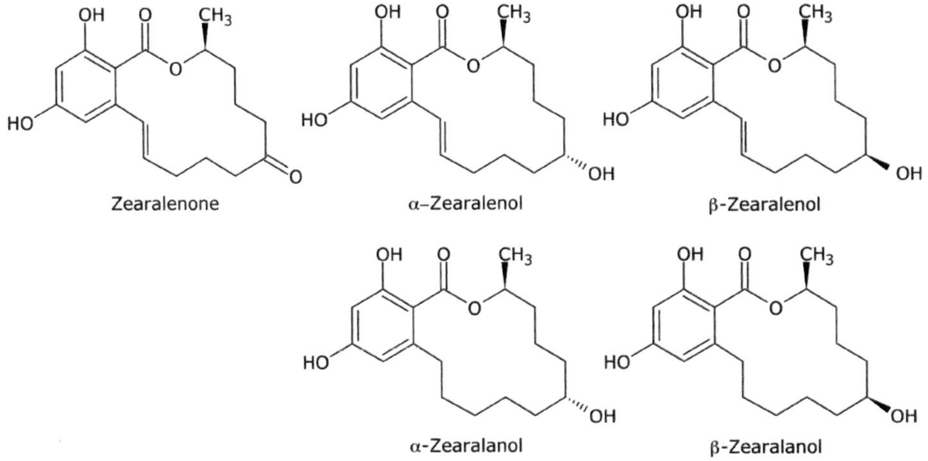

Figure 1.12 *– Chemical structures of zearalenone and its derivatives: α-zearalenol, β-zearalenol, α-zearalanol and β-zearalanol.*

trichothecenes and FUM (Zinedine *et al.*, 2007). Chemically, ZEN is a phenolic resorcylic acid lactone (Zinedine *et al.*, 2007). Derivatives of ZEN are α-zearalenol, β-zearalenol, α-zearalanol and β-zearalanol.

1.9.5.2 Metabolism/mechanism of action

Zearalenone is a non-steroidal estrogenic mycotoxin, which is often involved in reproductive disorders and hyperestrogenicity in farm animals. The estrogenic effects are based on the structural similarity between ZEN and estradiol. Estradiol is the most important female sex hormone in the group of estrogens and ZEN competitively binds to estrogen receptors (Vila-Donat *et al.*, 2018). The mycotoxin passively enters the cell via the cell membrane and binds to the estrogen receptor (Figure 1.13). This complex is transferred into the nucleus where it binds to specific nuclear receptors. Subsequently it generates estrogenic responses by inducing the transcription of genes normally expressed in case of receptor–estrogen complex binding (Riley and Norred, 1996). Estrogenicity applies to all ZEN forms, but the reduced form of ZEN, α-zearalenol, has increased estrogenic effects. These four derivatives (mentioned before) and additionally zearalanone can be found in infected plants, but at much lower concentrations than ZEN (Zinedine *et al.*, 2007).

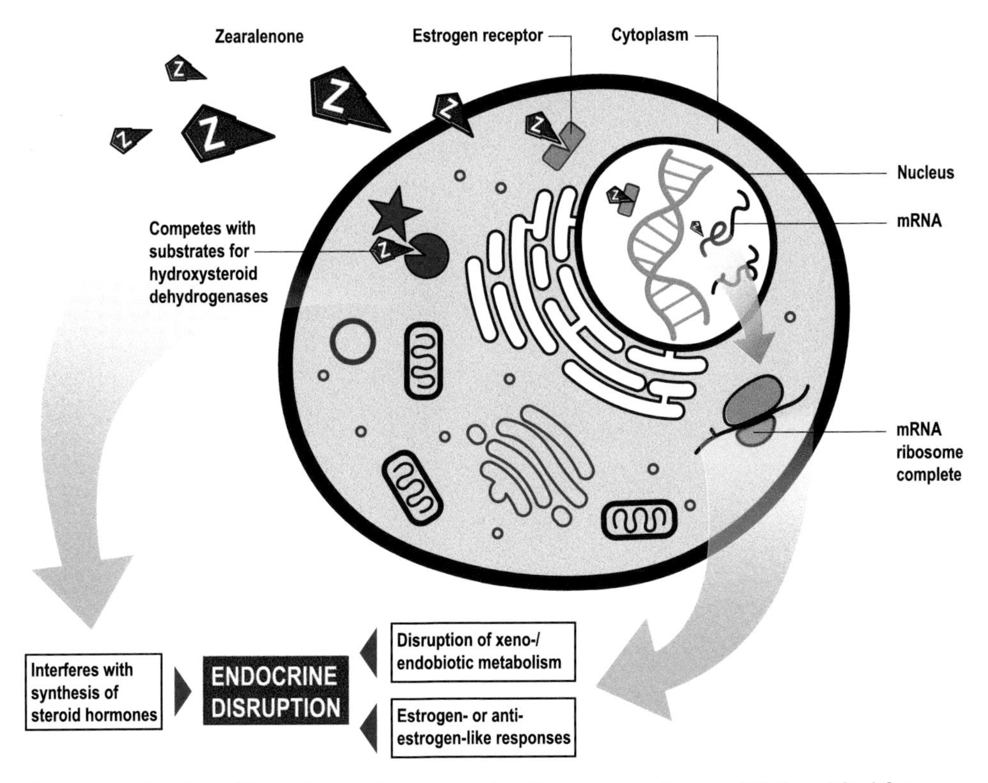

***Figure 1.13** – Simplified view of the mode of action of zearalenone (Z) (modified from Riley and Norred, 1996 and reviewed in Fink-Gremmels and Malekinejad, 2007; Zinedine* et al., *2007).*

The biotransformation of ZEN takes place in two major pathways (Minervini and Dell'Aquila, 2008):

- hydroxylation: formation of α-zearalenol and β-zearalenol, this reaction is assumed to be catalyzed by 3α- and 3β-hydroxysteroid dehydrogenases
- conjugation of ZEN and its metabolites with glucuronic acid catalyzed by uridine diphosphate glucuronyl transferases.

In animals, ZEN is metabolized to a high degree to α-zearalenol, β-zearalenol, α-zearalanol and β-zearalanol (Zinedine *et al.*, 2007). All four derivatives are estrogenic. Importantly, α–zearalenol shows higher estrogenicicty than ZEN. Several studies indicated that there are differences in biotransformation of ZEN in various species (Malekinejad *et al.*, 2006).

1.9.5.3 Exposure and absorption into the organism

Zearalenone is rapidly absorbed and metabolized in intestinal cells. In pigs, ZEN and its metabolites show an extensive biliary excretion and enterohepatic cycling (Biehl *et al.*, 1993). Studies investigating the carry-over of ZEN into meat and other edible tissues indicated that there is only limited tissue deposition of this mycotoxin. This is probably due to its rapid biotransformation. Transfer of ZEN and its major metabolites into serum was not detected after an administration of 56 μg ZEN per kg feed (Goyarts *et al.*, 2007). Transfer of ZEN to milk was detected when cows were fed high doses. When doses of 50 or 165 mg ZEN (equivalent to 0.1 and 0.33 mg kg^{-1} bodyweight) were fed for 21 days, no residues could be found in the milk (Zinedine *et al.*, 2007). Zearalenone detection in commercial eggs has not been reported (Zinedine *et al.*, 2007).

1.9.5.4 Toxicity

IARC concluded that ZEN's carcinogenicity to humans is not classifiable (Group 3) (IARC, 1993; Ostry *et al.*, 2016). Nevertheless, the question of a connection between breast cancer and ZEN has been raised. The connection is proposed due to epidemiological data and findings that ZEN can interact with estrogenic receptors in mammary glands (Marin *et al.*, 2013).

In animals, ZEN shows a relatively low acute toxicity and the oral LD_{50} values are between 2,000 and 20,000 mg kg^{-1} bodyweight. However, there are severe chronic effects. Effects differ between species (Table 1.10). The toxin has been shown to provoke hepatotoxic, genotoxic and hematotoxic effects (Zinedine *et al.*, 2007).

Most effects are caused by the estrogenic effects of ZEN and its intermediates. In female animals, during/after pregnancy decreased fetal weight and survival of embryos as well as retention or absence of milk were observed (Marin *et al.*, 2013; Vila-Donat *et al.*, 2018; Zinedine *et al.*, 2007). Furthermore, the uterus tissue and morphology can be altered, decreased fertility, reduced litter size, swollen red vulvas and vaginal or rectal prolapse are documented (Iheshihulor, *et al.*, 2011; Zinedine *et al.*, 2007). For example, in zebrafish ZEN was shown to reduce the frequency of spawn and to provoke abnormalities in larvae (Schwartz *et al.*, 2010, 2013) (see Chapter 3). In male animals, reduced libido, lower spermatogenesis, decreased size of testicles, reduced testosterone levels and feminization were reported (Iheshihulor *et al.*, 2011; Marin *et al.*, 2013; Zinedine *et al.*, 2007). Immunosuppression was observed in both, males and females (Zinedine *et al.*, 2007).

Table 1.10 **–** *Main systems affected by zearalenone. (All animal species are described in this table. For the effects of zearalenone on aquaculture species, please see Chapter 3.)*

Affected system	Effects/signs/symptoms
Digestive system	Gastro-intestinal effects (diarrhea), rectal prolapse
Reproduction system	Reproductive effects (decreased fetal weight, survival of embryos, retention/absence of milk, decreased fertility, reduced litter size, enlargement of mammary glands, reddening and swelling of vulva, atrophy of ovaries, uterus hypertrophy; feminization, impaired semen quality, testicular atrophy, swollen prepuce, lower spermatogenesis, reduced testosterone levels)
Genes/Gene expression	Teratogenic effects (splay legs)
Immune system	Immunosuppression

1.9.6 Ergot alkaloids

1.9.6.1 General aspects

The term ergot alkaloid refers to a diverse group of approximately 40 different toxins formed by *Claviceps* spp., which occur on grains, such as triticale, corn, wheat, barley, oats, millet, sorghum and rice, as well as on some grasses. Toxic alkaloids are also produced by fungal endophytes, such as *Neotyphodium* spp. They colonize vegetative and reproductive tissues in perennial rye grass and tall fescue (EFSA, 2005a; Krska and Crews, 2008; Scott, 2009). Ergot alkaloids constitute the largest known group of nitrogenous fungal metabolites and all have a common structure, a tetracyclic ergoline ring system of lysergic acid. Clavines are the simplest Ergots containing only an ergoline ring system (Figure 1.14 – top), whereas the ergopetides have an additional peptide moiety linked to the basic structure (Figure 1.14 – bottom).

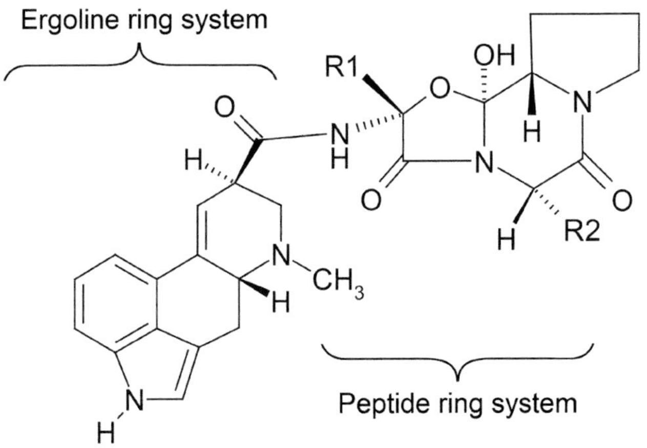

Figure 1.14 **–** *Structure of ergopeptines (Krska and Crews, 2008).*

The main groups of natural Ergots (CAST, 2003; Panaccione, 2005):

- clavines – e.g. agroclavine
- lysergic acids
- lysergic acid amides – e.g. ergonovine (ergometrine, ergobasine), ergine
- ergopeptines – e.g. ergovaline, ergotamine, ergocornine, ergocristine, ergosine, ergocryptine.

The amount and pattern of the different toxins vary between fungal strains, host plants and climatic and geographical conditions (Hafner *et al.*, 2008). Ergot alkaloids appear as colorless crystals that are soluble in various organic solvents, but insoluble or only slightly soluble in water (EFSA, 2005a).

1.9.6.2 Absorption/excretion/residues

In general, little carry-over of Ergots into animal tissue has been reported. When for instance a diet of 4% ergot (containing ergopeptine alkaloids) was fed to pigs, there was 90% absorption but no evidence of alkaloids in tissues (Scott, 2009).

1.9.6.3 Mechanism of action

The biological activity of Ergots in animal systems is mainly due to structural similarities of the ergoline ring structure to the neurotransmitters noradrenaline, adrenaline, dopamine and serotonin (Berde, 1980; Weber, 1980). Owing to this structural similarity, many Ergots can bind to neurotransmitter receptors and elicit effects such as a decrease in serum prolactin or vasoconstriction. Ergot alkaloids constitute a very diverse class of chemical compounds with widely different toxicological targets and activities, and potential routes of elimination and rates of clearance (Strickland *et al.*, 2011).

1.9.6.4 Toxicity

Animals can be exposed to complex mixtures of alkaloids in many typical animal agriculture production systems. The kinds of alkaloids present and their levels can vary widely, depending on the fungal strain, the host plant and environmental conditions. Ergot alkaloid toxicoses in livestock are widespread and result in disruption of several physiological systems (reproduction, growth, cardiovascular) (Strickland *et al.*, 2011) (Table 1.11). Ergot alkaloids exert toxic effects on all animal species, and the most prominent toxic signs can be attributed to the interaction of Ergots with adrenergic, serotinergic and dopaminergic receptors (EFSA, 2005a).

1.10 Regulations for mycotoxin contamination

Rui A. GONÇALVES

Mycotoxins are responsible for significant economic losses due to the direct spoilage of feed products (CAST, 2003; Shane and Eaton, 1994; Vasanthi and Bhat, 1998), but can also cause diseases when consumed by humans, livestock or aquaculture species (Zain, 2011). Despite having been identified as categorically undesirable for most aquaculture species, their occurrence, at least under field conditions, is not completely preventable even when using good manufacturing practices. Despite recent efforts to document mycotoxin occurrence in aquaculture feeds, we are still far from having a good overview of this topic. One of the biggest challenges is the large number of aquaculture-farmed species. Different

Table 1.11 – *Main systems affected by ergot alkaloids. (All animal species are described in this table. For the effects of ergot alkaloids on aquaculture species, please see Chapter 3.)*

Affected system	Effects/signs/symptoms
Circulatory system	Vasoconstriction symptoms (elevated body temperatures, increased respiration rate; vasoconstriction in extremities can lead to the loss of a limb), gangrenous changes in tissue of feet, tail and ear, reduced serum prolactin
Immune system	Immunosuppression (decreased resistance to environmental and microbial stressors; increased susceptibility to diseases)
Digestive system	Gastro-intestinal effects (reduced body weight gain, feed refusal, diarrhea)
Reproductive system	Decreased breeding efficiency (lower conception rates, decreased survival rate of offspring, decreased piglet birth weight, abortions)
Nervous system	Neurotoxic effects (convulsions, hallucination, anorexia), lameness
Skin	Dermotoxicity (oral and dermal lesions; necrosis), rough hair coat

species, even those belonging to the same trophic level, are commonly fed with different raw materials according to local availability and price. It is therefore impossible to extrapolate occurrence results from one species to another. A further challenge is that regulation of mycotoxin levels in feed is not harmonized worldwide, which would be beneficial due to the globalization of the trade of plant commodities and would increase the awareness of mycotoxins entering the feed supply chain.

1.10.1 Worldwide regulation of mycotoxin levels in aquafeed with focus on the EU and the USA

Mycotoxin levels in feed are regulated in several countries. However, the maximum acceptable limits vary greatly from country to country (FAO, 2004). The EU harmonized maximum levels and guidance values for mycotoxins in feed among its member states. Maximum levels for Afla and for rye ergots (*Claviceps purpurea*) have been set (Commission Regulation [EU] No. 574/2011, which amends Annex I to Directive 2002/32/EC of the European Parliament and of the Council on undesirable substances in animal feed; Commission Regulation [EC] No. 1881/2006 defining maximum levels for certain contaminants in foodstuffs). Nevertheless, the regulatory agencies of each country are allowed to set higher standard rules. For DON, ZEN, OTA and FUM, guidance levels have been set (Commission Recommendation 2006/576/EC). Unlike regulations, directives and decisions, recommendations are not legally binding. For T-2 and HT-2 only indicative levels are defined (Commission Recommendation 2013/165). Above these values further investigations should be performed, especially when found repeatedly. Nevertheless, these values are not levels for food and feed safety.

In the USA, the Center for Veterinary Medicine of the Food and Drug Administration (FDA) deals with mycotoxins in food and feed. There are action, guidance and advisory levels for Afla, FUM and DON, respectively (http://www.fda.gov; accessed 17 September 2018). Action Levels mean that the FDA wants to define a precise level of contamination. At this level, the FDA is “prepared to take regulatory action". For OTA and ZEN, the FDA has issued neither action, guidance nor advisory levels. These two mycotoxins are handled on a case-by-case basis.

The awareness of mycotoxin-related issues in the aquaculture industry has only recently increased, mainly due to the reduction of fish meal use and increased inclusion levels of plant meals in aquafeeds. Therefore, contrary to livestock species, specific recommendations for mycotoxin maximum levels in aquaculture feed do not yet exist, except for FUM, which are covered by guidance levels in the EU and the USA. In the EU, aquaculture feeds are covered by maximum and guidance levels generally applicable to feedstuffs, which are in many cases higher than levels set for specific livestock species (Table 1.12). While this might be sufficient for some species, e.g. channel catfish, which seem to be resistant to different types of mycotoxins, it is certainly not sufficient for other species such as white leg shrimp, which are extremely sensitive to several mycotoxins. For example, the only guidance value specifically addressing aquaculture species (FUM < 10 mg/kg) might be safe for certain but not all species (for example, turbot, Gonçalves *et al.*, in press-a). The situation is similar for aquaculture feeds in the USA (Table 1.13).

***Table 1.12** – Regulatory limits and guidance values for mycotoxins in animal feed in the EU applicable to aquaculture feedstuffs. Iceland, Norway and Liechtenstein are following EU legislation for aflatoxins and ergot alkaloids. Limits in µg/kg (ppb) or mg/kg (ppm) relative to a feeding stuff with a moisture content of 12%*

Sources: Commission Regulation (EC) No 1881/2006; Commission Regulation (EC) No 1137/2015; Commission Regulation (EC) No 1126/2007; Commission Regulation (EC) No 239/2016; Commission Regulation (EC) No 105/2010; Directive 2002/32/EC; Commission Regulation (EC) No 165/2010; Commission Recommendation 2006/576/EU; Commission Recommendation 2013/165/EU; Commission Recommendation 2013/627/EU; Commission Regulation (EC) No 212/2014; Commission Recommendation 2016/1319/EU.

Maximum levels	
Aflatoxins	
Commodity	**Limit (µg kg^{-1})**
All feed materials	20
Complementary and complete feed	10
Guidance values	
Fumonisins (B_1 + B_2)	
Commodity	**Limit (µg kg^{-1})**
Maize and maize products	60,000
Complementary and complete Feed stuffs for fish	10,000
Ochratoxin A	
Commodity	**Limit (µg kg^{-1})**
Cereals and cereal products	250

***Table 1.12** – Contd.*

Deoxynivalenol	
Commodity	**Limit (μg kg⁻¹)**
Cereals and cereal products with the exception of maize by-products	8,000
Maize by-products	12,000
Complementary and complete Feed stuffs	5,000
Zearalenone	
Commodity	**Limit (μg kg⁻¹)**
Cereals and cereal products with the exception of maize by-products	2,000
Maize by-products	3,000

***Table 1.13** – Action, guidance and advisory levels for mycotoxins in animal feed in the USA applicable to aquaculture feedstuffs. Concentrations in μg/kg (ppb). Feed stuffs and proportion in the diet based on dry weight basis.*

Action levels	
Aflatoxins	
Animals	**Limit (μg kg⁻¹)**
For immature animals: corn, peanut products, other animal feeds and feed ingredients, excluding cottonseed meal	20
Dairy animals, animal species or uses not specified, or when the intended use is unknown: corn, peanut products, cottonseed meal, and other animal feeds and feed ingredients	20
Guidance levels	
Fumonisins ($B_1 + B_2B_1 + B_2$)	
Animals	**Limit (μg kg⁻¹)**
Swine and catfish: Corn and corn by-products (no more than 50% of diet)/Complete ration	20,000/10,000
All other species or classes of livestock and pet animals: Corn and corn by-products (no more than 50% of diet)/Complete ration	10,000/5,000
Advisory levels	
Deoxynivalenol	
	Limit (μg kg⁻¹)
All other animals: grains and grain byproducts (not to exceed 40% of the diet)	5,000

Please see at the end of this chapter Tables 1.14 and 1.15 for EU regulations for all animal feed and Table 1.16 for all FDA regulations.

In general, there is an urgent need for more studies on the impact of mycotoxins and other fungal metabolites on aquaculture species in order to establish sensitive limits for, at least, the most commercially important species. Moreover, regulatory limits and guidance values for mycotoxins in feed should take into account particular aquaculture species or certain production sectors, for example, shrimp production. Furthermore, mycotoxin regulatory limits and guidance values need to consider animal health and welfare, as well as human health (Chapter 3).

1.10.2 Bioaccumulation of mycotoxins in aquaculture species

Bioaccumulation of mycotoxins originating from feed in edible tissues might represent a direct risk to human health (CAST, 2003). Mycotoxin bioaccumulation in livestock is well investigated (Leeman *et al.*, 2007; Völkel *et al.*, 2011) and the risk to humans is currently being evaluated by the European Food Safety Authority (EFSA) for several mycotoxins (Afla, OTA, ZEN, DON, FUM, T-2 and HT-2). Bioaccumulation of mycotoxins in poultry, swine and cows is managed by direct regulation of mycotoxins in animal feed (EC, 2006; EFSA, 2004a, 2004b, 2004c, 2004d; 2005b; 2011b; 2013). While regulatory limits have been put in place for Afla, only guidance values are available for DON, OTA, FUM and ZEN (EC, 2006). These guidance values take into consideration occurrence and transfer factors for livestock species, but not for aquaculture species. Currently, no regulations or guidelines exist in order to avoid deposition of mycotoxins in edible tissues of farmed fish or shrimp, with the exception of FUM ($FB_1 + FB_2 = 10$ mg kg^{-1}; EC, 2006). Moreover, it is not taken into consideration that carry-over mechanisms in aquaculture-farmed species might be different from terrestrial livestock species. Generally, the possibility of mycotoxin bioaccumulation/biomagnification through the food chain due to the use of mycotoxin contaminated non-plant origin ingredients such as animal by-products (for example, shrimp head meal or chicken droppings or non-typical mycotoxin contaminated ingredients (for example, fishmeal)) are not considered (for more information, consult Gonçalves *et al.*, in press-a).

Recently, Goncalves *et al.* (in press-b), critically reviewed the occurrence of mycotoxins in aquafeeds as well as the possibility of carry-over of these mycotoxins. The authors highlighted that particular attention should be paid to aquaculture edible tissues and that regional guidance limits should be advised depending on local mycotoxin occurrence and the edible tissues consumed. Risk assessment of imported aquaculture foods needs to take into account the mycotoxin occurrence in feed in the region of origin, which is especially important for products imported from highly mycotoxin contaminated regions, or regions known to use potentially contaminated animal by-products.

Furthermore, the authors highlighted that the available carry-over studies indicate that deposition of mycotoxins in edible tissues of aquaculture species may be higher than in terrestrial species and it is therefore imprudent to assume the same transfer factors for aquaculture species as for livestock species.

1.11 Legislation versus safe levels of mycotoxins

The development of legislation is crucial to limit the dietary exposure of animals to mycotoxins. However, as previously mentioned, animals are often confronted with other challenges that increase their susceptibility to these hazardous substances. Furthermore, co-contamination of mycotoxins increases the risk for animal health. Consequently, animals in the field can already be negatively affected by mycotoxin levels that are lower than maximum levels set by the legislation or concentrations shown to be hazardous in scientific studies. Low levels of mycotoxin contamination might also be important to consider in terms of possible bioaccumulation of mycotoxins (see above).

The negative impact of the toxins depends not only on the level and the type of contamination, but also on the general health status of the animal and on environmental conditions. Farm animals with symptoms typical for chronic mycotoxicoses have been observed, although the levels of mycotoxins did not exceed guideline levels (Vila-Donat *et al.*, 2018). All levels of mycotoxins should be considered as unsafe and increased levels are associated with increased risks to animal health. Even low levels of dietary mycotoxins can have a detrimental effect on the immune system and this is a hindrance for optimum performance.

1.12 Acknowledgements

A special thank you to Dr. Roman Labuda and Dr. Georg Häubl for carefully revising this chapter and for sharing their knowledge of mycology, microbiology and chemistry with us.

Parts of this text were originally published in *Guide to Mycotoxins* (Krska *et al.*, 2012) and written by Karin Naehrer, updated by Anneliese Mueller.

***Table 1.14** – Regulatory limits and guidance values for mycotoxins in animal feed in the EU. Iceland, Norway and Liechtenstein are following EU legislation for Aflatoxins and Ergot alkaloids. Limits in µg/kg (ppb) or mg/kg (ppm) relative to a feeding stuff with a moisture content of 12%*

Maximum levels	
Aflatoxins	
Commodity	**Limit (µg kg⁻¹)**
All feed materials	20
Complementary and complete feed stuffs	10
With the exception of:	
Compound feed for dairy cattle and calves, dairy sheep and lambs, dairy goats and kids, piglets and young poultry animals	5
Compound feed for cattle (except dairy cattle and calves), sheep (except dairy sheep and lambs), goats (except dairy goats and kids), pigs (except piglets) and poultry (except young animals)	20
Ergot alkaloids/Rye Ergot (*Claviceps purpurea*)	**Limit (mg kg⁻¹)**
Feed materials and compound feed containing unground cereals	1,000

(Continued)

Table 1.14 – *Contd.*

Guidance values	
Fumonisins ($B_1 + B_2B_1 + B_2$)	
Commodity	**Limit ($\mu g\ kg^{-1}$)**
Maize and maize products	60,000
Complementary and complete feed stuffs for:	
pigs, horses (*Equidae*), rabbits and pet animals	5,000
fish	10,000
poultry, calves (< 4 months), lambs and kids	20,000
adult ruminants (> 4 months) and mink	50,000
Ochratoxin A	
Commodity	**Limit ($\mu g\ kg^{-1}$)**
Cereals and cereal products	250
Complementary and complete feed stuffs for:	
cats and dogs	10
pigs	50
poultry	100
Deoxynivalenol	
Commodity	**Limit ($\mu g\ kg^{-1}$)**
Cereals and cereal products with the exception of maize by-products	8,000
Maize by-products	12,000
Complementary and complete feed stuffs	5,000
With the exception of:	
Complementary and complete feed stuffs for pigs	900
Complementary and complete feed stuffs for calves (< 4 months), lambs, kids and dogs	2,000
Zearalenone	
Commodity	**Limit ($\mu g\ kg^{-1}$)**
Cereals and cereal products with the exception of maize by-products	2,000
Maize by-products	3,000
Complementary and complete Feed stuffs for:	
piglets and gilts (young sows)	100

Table 1.14 – Contd.

Zearalenone	
Commodity	**Limit (μg kg⁻¹)**
sows and fattening pigs	250
calves, dairy cattle, sheep (including lamb) and goats (including kids)	500

Table 1.15 *– Indicative levels for the sum of T-2 and HT-2 in the EU, from which onwards/above which investigations should be performed, certainly in case of repetitive findings. Levels in μg/kg (ppb) relative to a feed with a moisture content of 12%*

Indicative levels	
T-2 and HT-2	
Commodity	**Limit (μg kg⁻¹)**
Cereal products for feed and compound feed	
oat milling products (husks)	2,000
other cereal products	500
compound feed, with the exception of feed for cats/compound feed for cats	250/50

Table 1.16 *– Action, guidance and advisory levels for mycotoxins in animal feed in the USA. Concentrations in μg/kg (ppb). Feed stuffs and proportion in the diet based on dry weight basis.*

Action levels	
Aflatoxins	
Animals:	**Limit (μg kg⁻¹)**
Finishing (i.e., feedlot) beef cattle: Corn and peanut products	300
Beef cattle, swine, or poultry: Cottonseed meal	300
Finishing swine of 100 pounds (45 kg) or heavier: Corn and peanut products	200
Breeding beef cattle, breeding swine, or mature poultry: Corn and peanut products	100

(Continued)

***Table 1.16** – Contd.*

Action levels	
Aflatoxins	
Animals:	**Limit ($\mu g\ kg^{-1}$)**
For immature animals: Corn, peanut products, other animal feeds and feed ingredients, excluding cottonseed meal	20
Dairy animals, animal species or uses not specified, or when the intended use is unknown: Corn, peanut products, cottonseed meal, and other animal feeds and feed ingredients	20
Guidance levels	
Fumonisins ($B_1 + B_2B_1 + B_2$)	
Animals:	**Limit ($\mu g\ kg^{-1}$)**
Equids and rabbits: Corn and corn by-products (no more than 20% of diet)/Complete ration	5,000/1,000
Swine and catfish: Corn and corn by-products (no more than 50% of diet)/Complete ration	20,000/10,000
Breeding ruminants breeding poultry and breading mink (incl. lactating cows, laying hens): Corn and corn by-products (no more than 50% of diet)/Complete ration	30,000/15,000
Ruminants ≥ 3 months old raised for slaughter and mink raised for pelt production: Corn and corn by-products (no more than 50% of diet)/Complete ration	60,000/30,000
Poultry being raised for slaughter: Corn and corn by-products (no more than 50% of diet)/Complete ration	100,000/50,000
All other species or classes of livestock and pet animals: Corn and corn by-products (no more than 50% of diet)/Complete ration	10,000/5,000
Advisory levels	
Deoxynivalenol	
Animals:	**Limit ($\mu g\ kg^{-1}$)**
Ruminating beef and feedlot cattle > 4 months: Grains and grain byproducts on an 88% dry matter basis	10,000
Ruminating dairy cattle > 4 months: Grains and grain byproducts on an 88% dry matter basis (not to exceed 50% of diet)	10,000

Table 1.16 – *Contd.*

Advisory levels	
Deoxynivalenol	
Animals:	**Limit (μg kg^{-1})**
Ruminating beef and feedlot cattle > 4 months, and ruminating dairy cattle > 4 months: Distillers grains, brewer's grains, gluten feeds, and gluten meals on an 88% dry matter basis	30,000
Chickens: Grains and grain by-products (not to exceed 50% of the diet)	10,000
Swine: Grains and grain byproducts on an 88% dry matter basis (not to exceed 20% of the diet)	5,000
All other animals: Grains and grain byproducts (not to exceed 40% of the diet)	5,000

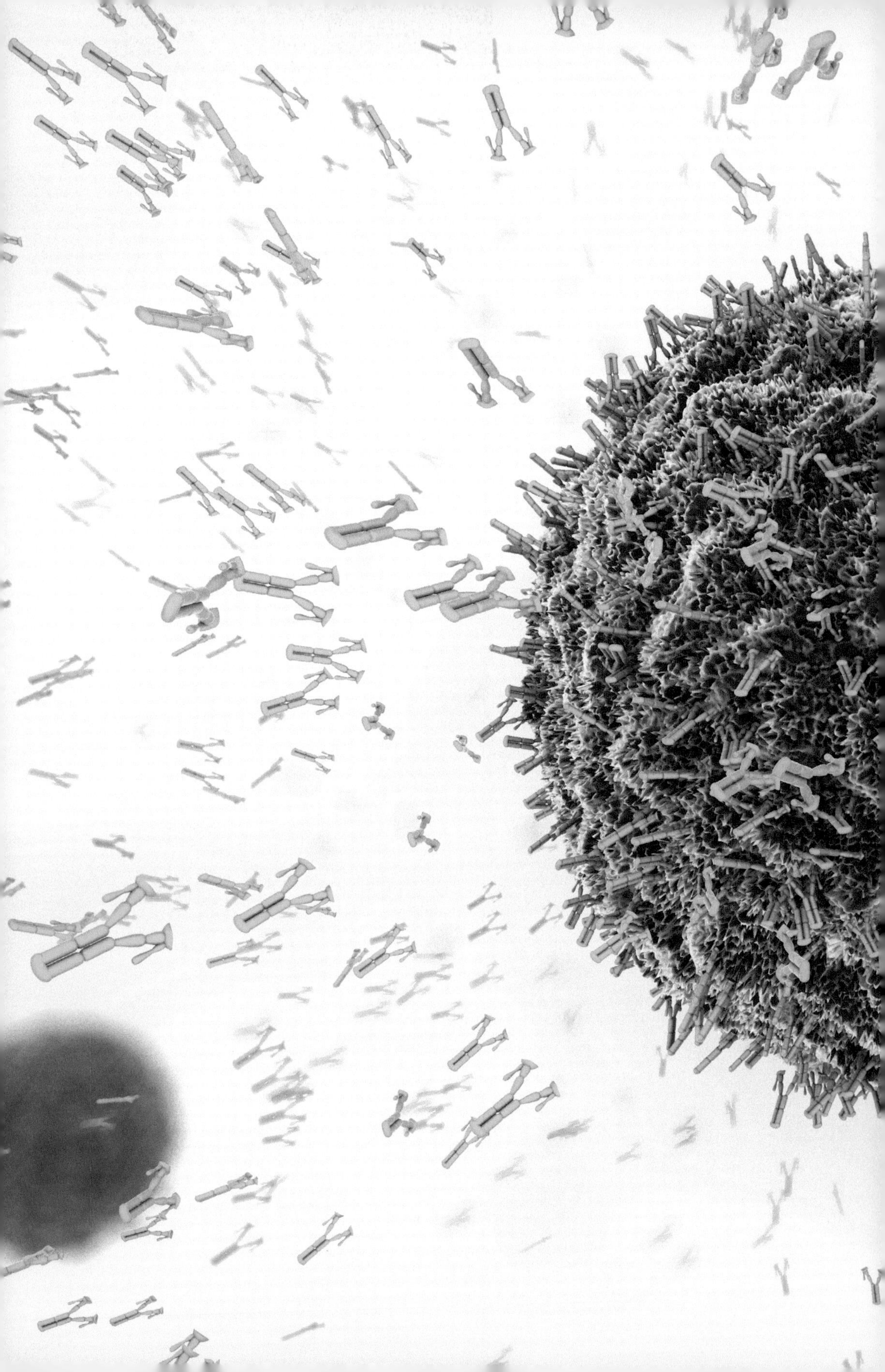

02

Aquatic species defense mechanisms

By Michele MUCCIO

2. Aquatic species defense mechanisms

By Michele MUCCIO

2.1 Vertebrates immune system

Immunity is defined as the resistance to diseases, with particular emphasis on infectious diseases. The whole of cells, tissues and molecules that mediate this resistance is known as the immune system, and the sequence of reactions that take place following the encounter of these components with infective entities is called immune response. The immune system of fish is very complex and well developed, sharing many features with mammals, as the result of millions of years of evolution (Foey and Picchietti, 2014).

Fish live in an environment that is by nature rich in pathogens and antigens, thus parasites, viruses and bacteria constantly challenge the immune system. The challenge is even greater in aquaculture production, where a great number of animals live in close contact, in a confined space, with often a very limited recirculation of water. All these factors cause stress in the animals and render the immune system less effective (Foey and Picchietti, 2014).

Fish possess both an innate and an adaptive immunity. The innate immunity is the first line of defense against infections and, as the name suggests, it is constantly active, whereas the adaptive immunity develops slower as a response to different pathogens that invade the tissues. Therefore, the adaptive immunity is specific to the type of pathogen that attacks the host, and it is stronger and more effective. The innate and the adaptive immune systems do not work individually but are in constant cooperation (Figure 2.1). The adaptive compartment utilizes several cell types from the innate compartment and potentiates their efficacy. In fish, a third component deserves particular attention: the mucosal immune system, located in the gills and the gut (Foey and Picchietti, 2014).

2.1.1 Innate immunity

In the innate compartment, the first line of defense is represented by the epithelial barriers and by specialized cells that are secreting antimicrobial molecules (Abbas and Lichtman, 2003).

The epithelia are physical barriers and include the skin, the mucous membranes in the gastrointestinal tract (GIT) and the gills (Lieschke and Trede, 2009). Epithelial cells such as the ones composing the skin are able to activate the immune system in case a breach takes place. Epithelial cells also secrete antimicrobial molecules such as lysozyme and defensins to break down the cell wall of pathogens (Foey and Picchietti, 2014). If the skin barrier is compromised and the pathogens are able to cross it, they are exposed to different populations of immune cells that are described further in this chapter.

The mucus, together with commensal bacteria, provides an important physical barrier to prevent the entrance of pathogens. The mucus is produced in the gills, the gut and in the external surface of the skin (Uribe *et al.*, 2011). It is mainly composed of mucin and glycoproteins. The mucus layer is constantly removed and regenerated, it washes away pathogens and prevents them from attaching to the host. The mucus is also a lubricant; it helps the locomotion and the osmoregulation (Uribe *et al.*, 2011).

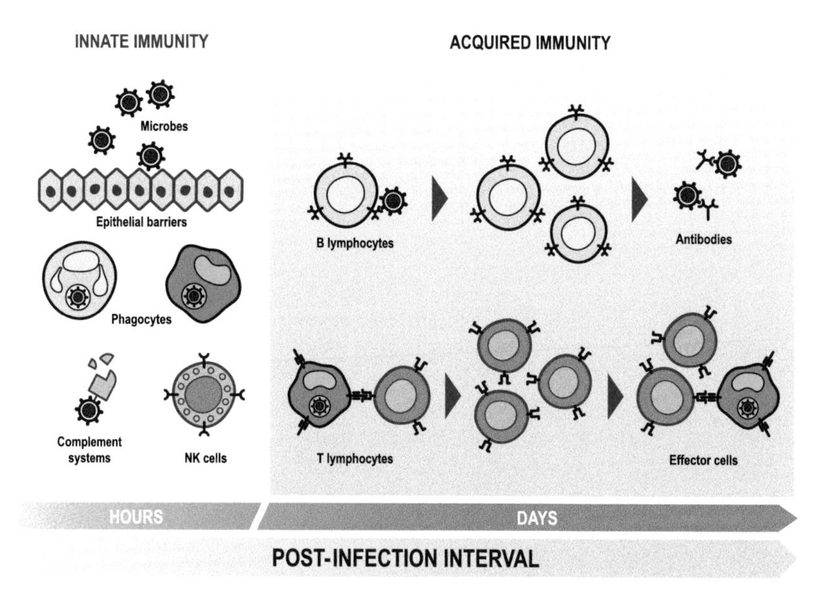

***Figure 2.1** – The innate immunity provides the first line of defense. Some components prevent the infection (e.g. epithelial barriers), whereas others eliminate microbes directly (e.g. phagocytes and natural killer (NK) cells). The acquired immune response appears later, and the response is mediated by lymphocytes and their products (adapted from Abbas and Lichtman, 2003).*

The cells of the innate compartment include macrophages and neutrophils involved in phagocytosis and secretion of innate inflammatory cytokines (Foey and Picchietti, 2014). The cells of innate immunity exclusively recognize microbes; they do not react to antimicrobial substances. The cells of the innate immunity do not potentiate after repeated exposure to the same pathogen. The whole of the cell lineages that constitute this complex system is described in the following paragraphs (Abbas and Lichtman, 2003).

2.1.2 Cells of the innate compartment

The innate immune system of fish includes neutrophils, monocytes, macrophages and granulocytes, such as basophils, eosinophils, mast cells and rodlet cells (Abbas and Lichtman, 2003). The function of neutrophils and macrophages is to recognize and phagocytize pathogens. In addition, they both can produce cytokines to alert other parts of the immune system. Cytokines are small proteins that are involved in cell signaling within the immune system. They contribute to stimulating the production of more phagocytes in response to an infection. Rodlet cells play a role in the defense against larger parasites such as helminths. Mast cells are composed of both acidophils and basophils, and are usually present in chronically inflamed tissues (Abbas and Lichtman, 2003).

In the case of an infection, the cells of the innate compartment are activated following recognition of conserved pathogen-associated molecular patterns (PAMPs). PAMPs are a broad array of molecules expressed by pathogens. The dedicated receptors for these molecules are called pattern recognition receptors (PRRs). Different classes of PRRs trigger different functions in the phagocytes, such as phagocytosis, chemotaxis (attraction of other immune cells to the infection site), inflammation and pathogen killing. A common PAMP is the lipopolysaccharide (LPS), an endotoxin produced by Gram-negative bacteria (Zou and Secombes, 2016).

The first cells that are recruited following an infection – especially in the presence of bacteria and fungi – are neutrophils. Neutrophils quickly migrate to the infection site where they phagocytize circulating microbes. Neutrophils have a relatively short life and die a few hours after their activation (Foey and Picchietti, 2014).

The function of monocytes is similar to that of neutrophils, but they are present in lower numbers. Monocytes reach the extravascular tissues and survive there for longer time, differentiating into cells known as macrophages. In addition to their phagocytic activity, macrophages produce cytokines that are recruiting other immune cells such as neutrophils and other monocytes to the infection site (Abbas and Lichtman, 2003).

Natural killer (NK) cells are a population of lymphocytes that recognize pathogens and produce interferon gamma (INF-γ), a soluble cytokine in charge of recruiting macrophages to the infection site, rendering the response more efficient. NK cells contain grains filled with proteases, apoptotic enzymes and perforins – a class of proteins that form holes in cell membranes. NK cells represent 10% of the total lymphocyte population in the bloodstream and peripheral tissues (Zou and Secombes, 2016).

2.1.2.1 How immune cells kill pathogens

Neutrophils and macrophages recognize microbes in the bloodstream thanks to the PAMP-PRR mechanism. Once microbes are recognized, neutrophils and macrophages proceed to phagocytize them. Phagocytosis is a process that involves the extension of the cell membrane around the microbe, followed by the welding of the two extremities behind the microbe. The cell membrane extension that surrounds the microbe is known as phagosome. The phagosome fuses together with other vesicles named lysosomes that contain hydrolytic enzymes and enzymes producing reactive oxygen and nitrogen species. This merger between the phagosome and lysosome creates a phagolysosome (Abbas and Lichtman, 2003). Once a microbe is trapped in this structure, the enzymes are released and the microbe is killed. Phagocytic oxidase converts molecular oxygen into the free radical superoxide, which is toxic to cells. Inducible nitric oxide synthase (iNOS) converts arginine into nitric oxide (NO), an important microbicide. Proteases and lysozyme degrade bacterial cell walls. During severe inflammation, these enzymes can be secreted extracellularly, damaging the host's tissues, too (Foey and Picchietti, 2014). The whole process is depicted in Figure 2.2.

2.1.2.2 Cytokines of innate immunity

Cytokines are soluble proteins that are secreted by macrophages and other immune cells upon exposure to microbes. These molecules are used for the communication between different classes of immune cells during the inflammation process. In the innate immunity, most of the cytokines are produced by activated macrophages following an infection process (Zou and Secombes, 2016). The most important cytokines, their functions and the cells producing them are listed in Table 2.1.

2.1.3 Adaptive immunity

Adaptive – or acquired – immunity provides a more specific and efficacious response to infections. It is able to develop memory, thus becoming quicker in reaction following multiple exposure to the same pathogen and it is able to adapt and strengthen following multiple exposure (Abbas and Lichtman, 2003). Compared with the innate counterpart, the adaptive compartment develops slowly and it needs an

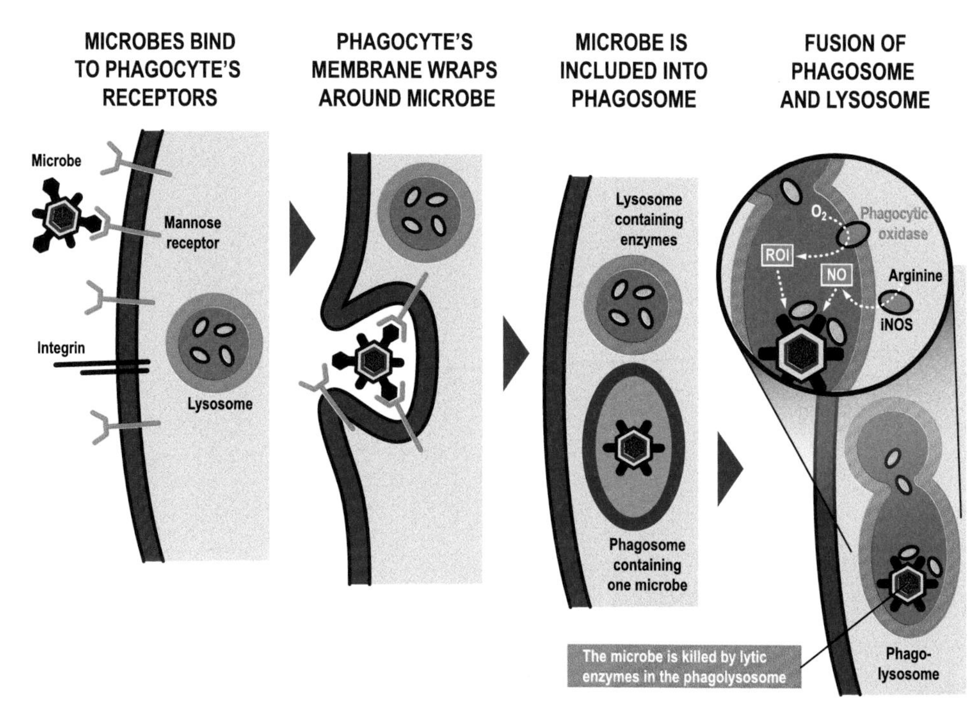

***Figure 2.2** – The membrane of macrophages and neutrophils exhibits different receptors that are able to bind microbes, facilitating the phagocytosis process. After the incorporation of microbes, phagosomes merge with lysosomes and microbes are killed by enzymes and toxic substances in the phagolysosome. Phagocytes can also release toxic substances in the extracellular environment to kill microbes. ROI: reactive oxygen intermediates (adapted from Abbas and Lichtman, 2003).*

activation, as it is not active at all times. Adaptive immunity is divided into humoral and cell-mediated immunity. The effector cells of the adaptive immune system are known as lymphocytes. These are subtypes of leukocytes or white blood cells and represent the effective part of the adaptive immune system. They include NK cells, B lymphocytes and T lymphocytes (Uribe *et al.*, 2011).

2.1.4 Humoral immunity

This compartment is in charge of eliminating extracellular pathogens and their toxins thanks to antibodies. Antibodies are large proteins produced by B lymphocytes, following their activation and differentiation into antibody-producing effective cells (plasma cells) (Magnadóttir, 2006). Antibodies are involved in the elimination of bacteria and viruses through the recognition of specific components called antigens that are produced by pathogens. The primary mode of action of antibodies is based on prevention of the infection. In fact, they prevent pathogens that are circulating in the bloodstream or the mucosa from colonizing tissues and organs (Magnadóttir, 2006).

The major limitation of antibodies is that they cannot eliminate pathogens that have entered cells. This function is taken care of by cell-mediated immunity through its main effector cells, the T lymphocytes.

***Table 2.1** – Different classes of cytokines, their main functions, targets and cells producing them (adapted from Abbas and Lichtman, 2003; Zou and Secombes, 2016).*

Cytokine	Main producer cell	Main cellular targets and functions
Tumor Necrosis Factor (*TNF*)	Macrophages, T lymphocytes	Endothelial cells: Activation (inflammation and coagulation) Neutrophils: activation Liver: acute phase protein synthesis Broad variety of cell types: apoptosis
Interleukin-1 (*IL-1*)	Macrophages, endothelial cells, some epithelial cells	Endothelial cells: activation Liver: acute phase protein synthesis
Chemokines	Macrophages, endothelial cells, T lymphocytes, fibroblasts, thrombocytes	Leucocytes: activation
Interleukin-12 (*IL-12*)	Macrophages and dendritic cells	T lymphocytes and natural killer (NK) cells: *INF-γ* synthesis, strengthening of cytotoxic activity T lymphocytes: differentiation
Interferon gamma (*INF-γ*)	NK cells, T lymphocytes	Activation of macrophages, promotion of some antibody response
Type 1 IFN (*IFN-α, IFN-β*)	*IFN-α*: macrophages *IFN-β*: fibroblasts	All cells: increase of PAMP (pathogen-associated molecular patterns) expression NK cells: activation
Interleukin-10 (*IL-10*)	Macrophages, T lymphocytes	Macrophages: inhibition of *IL-12* production, reduced class II major histocompatibility complex (MHC)
Interleukin-6 *(IL-6)*	Macrophages, endothelial cells, T lymphocytes	Liver: acute phase protein synthesis B lymphocytes: proliferation of antibody secreting cells
Interleukin-15 (*IL-15*)	Macrophages, other cells	NK cells: proliferation T lymphocytes: proliferation
Interleukin-18	Macrophages	NK cells and lymphocytes: synthesis of *INF-γ*

The mode of action of T lymphocytes works by either activating the macrophages, which are afterwards destroying the pathogens, or by killing the pathogens directly. T lymphocytes are specialized in recognizing antigens from intracellular pathogens, contrary to B lymphocytes, which are specialized in recognizing antigens from extracellular pathogens (Magnadóttir, 2006).

2.1.4.1 Specificity and memory of the humoral response

Specificity to structurally different antigens and memory of antigens that were previously encountered are the main characteristics of the acquired immunity. The immune system is able to discriminate between at least one billion different antigens (or portions of antigens), thanks to the presence of a huge, broadly diversified collection of receptors, known as immune repertoire (Lieschke and Trede, 2009). The total lymphocyte population is composed of many different subtypes ("clones"), expressing different receptors. The immune response is based on the ability of a specific antigen to bind a specific lymphocyte receptor (Magnadóttir, 2006).

The immune system responds with higher intensity and increased efficacy after repeated exposure to the same antigen (Abbas and Lichtman, 2003). The response to the first exposure to a certain antigen is known as primary response and it is mediated by cells known as virgin lymphocytes (they encounter the antigen for the first time). As these cells have never encountered the specific antigen before, they need more time to overcome the infection. If an antigen is encountered a second time, a secondary response will develop, which is quicker and more intense than the first. Secondary responses are the result of the activation of memory lymphocytes, a sub-type with a longer lifespan, produced following primary infection and important for generating a stronger immune response (Lieschke and Trede, 2009). With each subsequent exposure to the same antigen, the number of different B cell clones increases to generate a polyclonal response and a greater number of memory B cells as well (Figure 2.3) (Abbas and Lichtman 2003).

2.1.5 Cell-mediated immunity

The cell-mediated immunity is responsible for eliminating intracellular infections that cannot be reached by antibodies. The effector cells are the T lymphocytes and they are in charge of removing the infection (Magnadóttir, 2006; Uribe *et al.*, 2011).

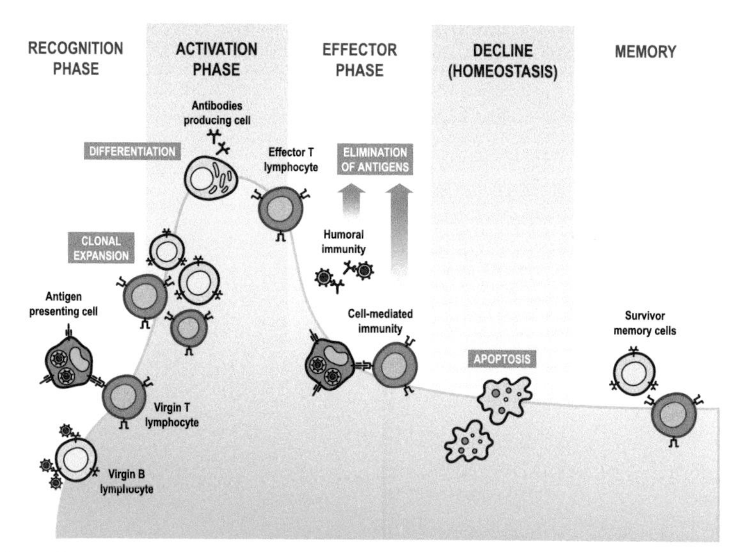

***Figure 2.3** – After antigen recognition, lymphocytes proliferate and differentiate into effector cells, which eliminate the specific antigen. After the elimination, the response decreases and the majority of activated lymphocytes goes into apoptosis, whereas some of the antigen-specific cells form the immunologic memory (adapted from Abbas and Lichtman, 2003).*

There are essentially two mechanisms that allow pathogens to cause an intracellular infection. First, certain bacteria are able to escape the bactericidal action of phagocytes. Second, certain viruses can bind specific receptors on some cell types, thus penetrating them and proliferating in the cytoplasm. Usually the infected cells do not possess any means to fight viruses (Zou and Secombes, 2016).

In the cell-mediated immunity, the T lymphocytes do not work alone against pathogens but need to cooperate with other cell lines such as phagocytes, host infected cells and B lymphocytes. This 'cooperation' is mediated by antibodies, which are produced by B lymphocytes, together with a sub-class of T lymphocytes known as $CD4^+$ T-helper cells (Abbas and Lichtman, 2003; Lieschke and Trede, 2009).

2.1.6 Cells of the acquired immunity

When it comes to elucidating the complex cell network of the acquired immunity, it is important to keep in mind the presence of two main groups of players: the lymphocytes specialized in recognizing antigens; and the effector cells that eliminate the pathogens (Magnadóttir, 2006).

Lymphocytes are the only cells equipped with specific receptors that can capture antigens and fight the infection. Although all lymphocytes look morphologically alike, there are profound functional differences. These cell groups are distinguished according to the type of proteins that are expressed on their membrane and that can be recognized by a specific group of antibodies (Abbas and Lichtman, 2003). These membrane proteins are classified with the 'CD (cluster of differentiation)' nomenclature. For each molecule, the initials CD are followed by a number that indicates the membrane proteins typical for a specific differentiation state, recognized by a specific cluster of antibodies (Magnadóttir, 2006; Uribe *et al.*, 2011).

Antigen presenting cells (APC) are part of the acquired immune system. These cells capture antigens from microbes and expose them on their own membrane to help the T lymphocytes to recognize them. APCs comprise dendritic cells, macrophages and B cells. A second class of APCs named professional APCs are able to express secondary signals such as specific microbial peptides, together with antigens (Lieschke and Trede, 2009).

Effector cells are in charge of physically eliminating the infection. This function is taken care of by lymphocytes and other leukocytes such as granulocytes and macrophages of the innate immunity (Lieschke and Trede, 2009).

2.1.7 Immune tissues in fish

The immune system of fish is composed of specialized organs and specialized cell types. Lymphocyte synthesis and development mainly takes place in lymphoid organs (Foey and Picchietti, 2014). Fish have two main lymphoid organs, the thymus and an analog of the bone marrow that is located in the head kidney and spleen. The evolution of the immune system in fish was linked to the primary route of infection of pathogens, which is via the mucosae. The mucosae protect fish from external stressors and contain different immunological tissues known as the mucosa-associated lymphoid tissue (MALT). The MALT can be divided into three immune compartments: skin-associated lymphoid tissue, gill-associated lymphoid tissue and gut-associated lymphoid tissue (GALT) (Foey and Picchietti, 2014).

2.1.7.1 The thymus

The thymus supports the development and maturation of T lymphocytes by providing an ideal microenvironment (Uribe *et al.*, 2011). The morphology of this organ varies between different species. In some species, it is possible to clearly identify the structure of the thymus, whereas it is difficult in other species where there are no clear zones delimitating it. In teleost fish, the thymus is located dorsally and it is covered by mucosal epithelium. Generally, the thymus consists of an internal lymphoid tissue containing the developing lymphocytes, which is surrounded by a capsule of epithelial tissue (Magnadóttir, 2006).

2.1.7.2 The head kidney and spleen

The kidneys are paired organs located in the body cavity on either side of the backbone and their primary function is excretion and regulation of the water balance within the fish (Zou and Secombes, 2016). The foremost part of the kidney lacks excretory capabilities and it is involved in immunologic processes such as hematopoiesis, antibody production and retention of antigens after vaccination. Owing to this, many authors suggested that the head kidney is the bone marrow analog, being the major source of B lymphocytes, which afterwards migrate to the spleen for activation (Foey and Picchietti, 2014).

The spleen is composed of white and red pulp, and the latter occupies most of the organ. The red pulp consists of a network of cells and blood vessels. It serves the purpose of an immune cell reservoir holding a heterogeneous population of cells, including macrophages and lymphocytes. The white pulp is usually less developed and the main function is plasma filtration, antibody synthesis and removal of antibody-coated bacteria (Abbas and Lichtman, 2003; Foey and Picchietti, 2014).

2.1.7.3 Skin-associated lymphoid tissue

Skin is a mechanical barrier against outside stressors and it secretes mucus. Several types of immune cells have been identified on the skin including leukocytes, macrophages and lymphocytes.

2.1.7.4 Gill-associated lymphoid tissue

The gills appear as a system of lamellae covered by respiratory epithelium. The gills are the main entry site for pathogens. Therefore, they contain several components of the innate and cell-mediated immunity to provide a first line of defense. Immune cells such as lymphocytes, macrophages, eosinophil granulocytes and antibody-secreting cells have been observed in the gill-associated lymphoid tissues of multiple species (Lieschke and Trede, 2009).

2.1.7.5 Gut-associated lymphoid tissue

In contrast to mammals, the fish gut is arranged in folds rather than villi and it lacks lymph nodes. GALT contains epithelial cells alongside different types of immune cells such as lymphoid cells, macrophages, eosinophils, neutrophils and a small number of B cells (Abbas and Lichtman, 2003). The gut itself deserves particular attention as it represents the first system that encounters foreign stressors that are contained in the feed. The integrity of the gut is crucial to prevent pathogens, toxins and other pathogenic entities from entering the body and causing more severe damage (Foey and Picchietti, 2014). Furthermore, a balanced gut microbiota is crucial to ensure gut health. The gut microbiota is the community of microbes present in the gut. Gut microbes may be symbiotic (i.e. beneficial to the host), commensal (i.e. non-harmful to the host) or pathogenic. Beneficial microbes support the digestive process by breaking down certain types of nutrients. Pathogenic microbes compete for the same substrates and can outgrow the beneficial counterparts, which is known as "dysbiosis". The presence of beneficial microbes that keep the growth of pathogenic microbes at bay is crucial for the overall wellbeing of the animal (Zou and Secombes, 2016).

2.1.8 Effects of mycotoxins on the immune system of fish

The current trend that sees a reduction of fishmeal in aquafeed formulation in favor of more cost-effective plant protein sources is raising the question whether mycotoxin contamination is an issue for aquaculture and what could be the consequences for producers. Answering this question is not easy due to the great variety of farmed species and production systems around the world. Furthermore, the existing literature on the topic is scarce and restricted to a few species, not all of commercial value (Gonçalves *et al.*, 2018c). When it comes to published studies on the effects of mycotoxins on fish immune system, the picture is even narrower, and only a few publications exist. We have seen that the immune system of fish is a complex mechanism that involves a large number of specialized cell types, molecules and tissues, and shares many conserved patterns with terrestrial vertebrates. In fact, when we speak about effector mechanisms that involve all the immune compartments that are physically in charge of eliminating the infection, we essentially speak of nearly the same ones that we see in terrestrial vertebrates. From the knowledge gained on terrestrial vertebrates we know that mycotoxins such as aflatoxins (Afla), fumonisins (FUM), trichothecenes, ochratoxins and zearalenone (ZEN) can have severe negative effects on cell proliferation and cell viability, and cause organ damage and carcinogenic effects.

Aflatoxins are some of the most powerful carcinogens. They are able to form adducts with the DNA with deleterious consequences for the organism (Marin *et al.*, 2013). Aflatoxins are currently the most feared mycotoxins in the industry due to their capacity to affect several species at very low concentrations – from 2.5 $\mu g\ kg^{-1}$ in carp (Gonçalves *et al.*, 2018c). Among the aquaculture scientific community, Afla are the most studied mycotoxins. Mycotoxins may affect immune organs or specific cell lines of the fish immune system. Sahoo and Mukherjee (2000) investigated the effects of aflatoxin B_1 (Afla B_1) administered via intraperitoneal injection on carp immune cells using an assay named NTB (nitroblue terazolium) which quantifies the amount of superoxide anion produced by phagocytic cells, as a weapon to kill invasive organisms. The authors observed that the production of superoxide anion was significantly decreased in phagocytic cells of fish treated with Afla B_1, thus rendering these cells less effective in killing pathogens (Sahoo and Mukherjee, 2000). Studies conducted on carp, reported a reduction of total plasmatic proteins (important carriers of hormones and immune system modulators) and a decreased production of lymphocytes. Both of these observations were attributed to the hepatotoxic effect of Afla. Aflatoxin-mediated pathological effects on liver and immune suppression were also observed in other species including tilapia, rainbow trout and hybrid sturgeon (Bailey *et al.*, 1994; Santacroce *et al.*, 2007).

The mode of action of trichothecenes such as deoxynivalenol (DON) and T-2 toxin (T-2) is based on the disruption of protein synthesis. The cytotoxic effects of trichothecenes are due to their ability to activate the apoptotic mechanism in target cells. Deoxynivalenol is most toxic to actively dividing cells, making the immune system one of its prime targets (Grenier and Applegate, 2013). Studies on the effects of DON on the immune system of carp revealed the capability of DON to partially impair the innate response (Pietsch *et al.*, 2014). The same study reported a reduction in size of erythrocytes after exposure to only 352 $\mu g\ kg^{-1}$ DON. Furthermore, fish fed low-doses of DON presented erythrocytes with increased activities of the antioxidant enzymes superoxide dismutase and catalase as a result of cytotoxic effects of DON (Anater *et al.*, 2016; Pietsch *et al.*, 2014). Salmonids such as Atlantic salmon and rainbow trout were reported to be sensitive to low levels of DON, however the effects on the immune system have not yet been investigated (Hooft *et al.*, 2011; Manning and Abbas, 2012).

Fumonisins are cytotoxic mycotoxins, with FB_1 being the most toxic and the most prevalent (Voss *et al.*, 2007). Fumonisins interfere with the metabolism of sphingolipids, important structural components of cell membranes. Without sphingolipids, the structure of the cell membrane is crucially compromised and consequently the cell is automatically terminated (Grenier *et al.*, 2013; Voss *et al.*, 2007). Rapidly dividing cells are the prime targets of FUM, suggesting synergistic interactions with trichothecenes. Manning and Abbas (2012) observed that fish exposed to mixtures of trichothecenes and FUM exhibited lower disease resistance and higher mortality when challenged with the pathogen *Edwardsiella ictaluri*. The authors noticed a markedly decreased antibody production as well (Manning and Abbas, 2012). According to the small number of studies that investigated the effects of trichothecenes and FUM on the immune system of fish, the main negative effects are comparable to that of terrestrial vertebrates (Manning and Abbas, 2012; Pietsch *et al.*, 2014). These effects are not species-specific but rather target conserved parts of the immune system such as different aspects of the humoral, cellular and innate compartments (Voss *et al.*, 2003). Some of the effects may concern changes in the expression of cell surface markers, which are relevant in the cell communication, antigen recognition and cytokine secretion (Anater *et al.*, 2016; Pepeljnjak *et al.*, 2003; Pietsch and Junge, 2016). Fumonisins have been shown to induce changes in the sphingolipid metabolism of channel catfish. A study that investigated the effects of FUM on the immune system of carp found these mycotoxins to cause multiple lesions to several internal organs, including main immunologic districts such as the head kidney and spleen. The same study observed a strong accumulation of rodlet cells – important components of the innate compartment – around damaged tissues, as a strong cell-mediated response to stress (Riley *et al.*, 2001).

Information on the effects of other mycotoxins such as ochratoxins and ZEN on the immune system of fish is scarce. One study on rainbow trout reported that these mycotoxins were able to damage the spleen and kidneys, important immunologic districts (Woźny *et al.*, 2012). Similar effects of ochratoxin A (OTA) were observed in a second study on rainbow trout (Gonçalves *et al.*, 2018c).

The immune system of fish presents many similarities with the one of terrestrial vertebrates, being a complex mechanism forged by millions of years of evolution. Low levels of mycotoxins are known to modulate the immune system of terrestrial vertebrates by targeting conserved components of the innate and adaptive immunity, either at cellular level, or by disrupting the activity of immunologic districts. Based on knowledge of deleterious effects of mycotoxins on the immune system of terrestrial vertebrates and a small number of studies performed in fish, we can speculate that mycotoxins might modulate the immune system of fish as well. They could represent a serious problem to aquaculture due to the increasing inclusion of plant protein sources in aquafeed formulation. Considering the vulnerability of the industry to diseases, especially in rural areas, further research is needed to elucidate the effect of these toxic compounds on the immune system of fish.

2.2 Invertebrates immune system

2.2.1 Introduction to the immune system of shrimp

The immune system of crustaceans, such as shrimp, can be considered as less sophisticated compared with the one of terrestrial vertebrates, since crustaceans only possess the innate compartment and have no adaptive immunity. The immune response is much simpler and it is the result of a network of interactions between several humoral cell groups (Söderhäll, 2010).

Crustaceans possess an open circulatory system where the hemolymph is the analog of blood in vertebrates (van de Braak, 2002). The hemolymph circulates through the body thanks to the heart, and gets in direct contact with the organs. The hemolymph is composed of a plasma fluid and it is the vehicle of distribution of nutrients, oxygen, hormones and effector cells. The hemolymph contains the hemocytes, the analogs of phagocytes in vertebrates. The hemocytes are involved in recognition and elimination of threats and have the function of coagulation as well (Vasquez *et al.*, 2008).

Despite the lack of an adaptive immune system, shrimp possess a series of effector mechanisms able to produce an immune response depending on the nature of the infection (Loker *et al.*, 2004). These effectors include the prophenoloxidase (proPO) system, phagocytosis and encapsulation (Söderhäll, 2010). The aim of this chapter is to provide a complete, yet concise overview of the different actors that play a role in the complex mechanism of immunity.

2.2.1.1 The proPO system

The proPO system is the analog of the complement system in higher vertebrates: the mechanism that renders phagocytes and antibodies more efficient in clearing the infection (usually by amplifying the release of cytokines) (Cooper and Alder, 2006). The proPO system plays a main role in the defense against large pathogens that cannot be phagocytized. To clear these kind of infections, the body uses melanization, intended as the encapsulation of a microbe within a melanin capsule (van de Braak, 2002). The proPO system is activated upon recognition of conserved signature molecules on the cell walls of pathogens, such as lipopolysaccharides and bacterial peptidoglycans (Vasquez *et al.*, 2008). The mechanism of recognition is based on the PRR-PAMP system already described in the fish immune system. Activation of the proPO system eventually leads to conversion of prophenoloxidase to its active form phenoloxidase. Phenoloxidase controls the process of melanization. During the formation of the capsule, free radicals are generated to actively kill the pathogen (Vasquez *et al.*, 2008).

2.2.1.2 Lectins

Lectins are crucial components of the immunity of invertebrates and are found in all crustaceans and in several unicellular organisms (Cooper *et al.*, 2006). Lectins are carbohydrate-binding proteins able to recognize carbohydrate residues that are specific to pathogens. They are mediators of phagocytosis, opsonization and agglutination. Lectins can be found free in the hemolymph and in the cytoplasm of hemocytes. The most diverse and well-studied lectins are the C-type lectins, so called due to their Ca^{2+} dependent carbohydrate-binding activity. Lectins are inducible and are produced in response to acute infection. Different organs such as the muscle, eyestalk, cuticle, hepatopancreas and hemocytes can carry out production of lectins. Different crustaceans show different expression sites of lectins; for instance, in *Litopenaeus vannamei* and *Penaeus monodon*, lectins are expressed only in the hepatopancreas and not in the hemocytes. Lectins in crustaceans share conserved structural homologies with the ones of higher vertebrates. These conserved residues helped researchers to identify lectins in crustaceans, while crustacean-specific residues helped in the identification of crustacean-specific pathogens.

2.2.1.3 Hemocytes and hematopoiesis

Hemocytes are the equivalent of white blood cells and carry out numerous functions including phagocytosis and recognition of pathogens. Shrimp have different types of hemocytes, such as hyaline cells (HC), semigranular cells (SGCs) and granular cells (GCs), which carry out different functions within the immune system (Loker *et al.*, 2004; Vasquez *et al.*, 2008). HC cells are mainly phagocytic, SGCs are mainly dedicated to pathogen recognition and the GC harbor granules that contain different immune factors such as the proPO activating system, cell adhesion and antimicrobial proteins (Loker *et al.*, 2004; Vasquez *et al.*, 2008).

Hematopoiesis occurs throughout the entire lifetime of the animal. It is regulated by factors that are secreted by the hemocytes themselves (Loker *et al.*, 2004; Söderhäll, 2010). Hematopoiesis occurs in specialized tissues called hematopoietic tissues (HPT), which consist of a pair of nodules in the dorsal side of the foregut or close to the antennal artery, depending on the species (Rusaini and Owens, 2010). The HPT produce hemocytes and release them into the circulation (van de Braak, 2002). They harbor several cell types, such as stem cells and precursors of hemocytes. Overall, there are five cell lineages, designated type 1 to type 5. Type 1 cells are characterized by large nuclei and a smaller cytoplasm resembling stem cells. Type 2 cells have larger nuclei and a larger cytoplasm. Type 1 and 2 cells are the main proliferating cells in the HPT. The function of cell types 3, 4 and 5 is not well understood, however, they are thought to be a kind of precursor of hemocytes (Vasquez *et al.*, 2008). In the HPT, the immature hemocytes go through a process of differentiation and specialization, such as the expression of the proPO that characterizes mature cells (Söderhäll, 2010). Final differentiation into active hemocytes occurs only at the time when the cells are released into the circulation. The proteins astakine 1 and astakine 2 have a function similar to cytokines. They potentiate the proliferation of hemocytes. Astakine 1 directly stimulates the proliferation of the HPT cells (precursors of hemocytes), whereas astakine 2 plays a role in the differentiation of granular cells (Loker *et al.*, 2004; Rusaini *et al.*, 2010; Söderhäll, 2010).

Only in penaeid shrimp, the lymphoid organ (LO) is the major phagocytic organ. It is responsible for phagocytosis of pathogens and foreign substances from the circulation (Rusaini *et al.*, 2010). Furthermore, it has bacteriostatic and antiviral defense functions. In males, the LO lies between the hepatopancreas and the stomach, whereas in females it is positioned between the ovary and the hepatopancreas. The size of the organ changes during the lifetime of the animal. It tends to increase as the animal grows. As reported by Rusaini *et al.* (2010), the size of this organ tends to increase in the post-larval stage, where the animal is more susceptible to viral and bacterial diseases.

2.2.1.4 Phagocytosis

Phagocytosis is carried out by hemocytes, which are normally present in the hemolymph as circulating cells. Additionally, other districts such as the surface of arterioles, hepatopancreas, gills and the LO in penaeids, can become phagocytic districts (Cooper *et al.*, 2006; Söderhäll, 2010). Phagocytes are specific to the kind of pathogen that is recognized. Some phagocytes recognize only Gram-negative bacteria, others are specific to Gram-positive bacteria. Cell factors and enzymes that are present in the hemolymph actively contribute to making phagocytosis more effective. Phagocytosis is triggered by the recognition of pathogen-specific residues by lectins and it is reinforced by the proPO system that mediates a more intense release of cytokines (Söderhäll, 2010). Other enzymes involved include peroxinectin, a protein that is released by hemocytes and serves as opsonin, that is, it directly promotes phagocytosis and encapsulation (Söderhäll, 2010).

2.2.1.5 Encapsulation

Encapsulation is a process that allows the elimination of large pathogens, such as helminths, fungal spores, large parasites or foreign particles (Cooper *et al.*, 2006; Söderhäll, 2010). The process involves the formation of a capsule around the pathogen that either kills it directly, or restricts its movements preventing it from causing damage. For capsule formation, different types of hemocytes aggregate around the particle thanks to the involvement of adhesive factors. A typical capsule is composed of several layers of hemocytes (ranging from 5 to 30 depending on the type of response) without any intercellular space, as the aim is to fully isolate the foreign compound from the host's body (Vasquez *et al.*, 2008).

2.2.1.6 Clottable proteins

Clottable proteins prevent the loss of hemolymph in case of injury. They have an immunologic function as well, as they help hemocytes to recognize and neutralize foreign components. In shrimp, the adhesive function is mediated by the enzyme transglutaminase, which is released by hemocytes in response to injury or invasion of the body by pathogens. Its expression is usually induced by third components such as the alteration of the Ca^{2+} balance. Clottable proteins can be found in the hemolymph and are not directly associated with the hemocytes (Loker *et al.*, 2004).

2.2.1.7 Antimicrobial proteins

Antimicrobial proteins are constitutively synthetized and stored in the hemocytes of all crustaceans. Their mode of action is based on the destabilization of microbial membranes mainly by the formation of pores. Antimicrobial proteins have the ability to modulate the immune system, thus rendering the immune response stronger. Some antimicrobial peptides were shown to be able to disrupt the metabolism of pathogens, for example, by interfering directly with DNA or RNA synthesis (Loker *et al.*, 2004; Vasquez *et al.*, 2008).

2.2.1.8 Antiviral factors

All crustaceans have a defense mechanism against viruses. Although the full mechanism is still unclear, it is known that some proteins, such as the antilypopolisaccharide factor (ALF), play an active role in this process (Söderhäll, 2010; van de Braak, 2002). Apoptosis following viral infection of the cell is an important antiviral response. Several proteins such as ribophorin I and caspases act as apoptosis regulators. Other antiviral proteins that are relevant for shrimp are penaeidins and crustins. Crustins contain a domain that resembles a protease inhibitor. These inhibitors target viral proteases and block the proteolytic cleavage of protein precursors that are necessary for the production of viral particles (Söderhäll, 2010; van de Braak, 2002).

2.2.2 The role of mycotoxins in the immune system of crustaceans

Since crustaceans do not possess an adaptive immune system, the immunologic function largely relies on the activity of hemocytes. The hepatopancreas – specifically the HPT – is the site of hemocyte production, maturation, specialization and release. If the health of the hepatopancreas is compromised, the whole immune system is compromised as well (Loker *et al.*, 2004).

When it comes to effects of mycotoxins in shrimp, studies are scarce and mostly focus on Afla. A recent work published by Zhao *et al.*, (2018) showed that dietary Afla disrupt the structure and function of the hepatopancreas and drastically increase the mortality of *L. vannamei*. The authors observed several histopathological changes such as detachment of the surface epithelia and cell necrosis. This was attributed to a negative effect of Afla on DNA synthesis and repair, as the expression levels of hepatopancreatic DNA polymerase subunits were markedly decreased. Aflatoxins had an effect on the metabolism of sphingolipids as well. These are essential structural components of cell membranes, and regulate metabolic functions, such as proliferation, differentiation and apoptosis. Aflatoxins interfere with the catalytic activity of the enzyme serine palmitoyltransferase (SPT), which plays an important role in the metabolism of sphingolipids and regulates cellular stress responses. When SPT is inactivated, toxic intermediate forms of sphingolipid metabolism accumulate in the cell, promoting apoptosis.

Another group of mycotoxins that is known to interfere with the metabolism of sphingolipids is FUM. These mycotoxins disrupt the pathway by inactivating the enzyme ceramide synthase, thus provoking an accumulation of free sphingosine (So) and sphinganine (Sa) in the cell and an increase in the ratio between free Sa and So, a commonly used biomarker for the effect of FUM. The process of sphingolipid biosynthesis is vital for cells and it is most important in rapidly dividing cells due to the greater metabolic turnover. Shrimp are very sensitive to FUM, with growth and feed conversion ratio (FCR) already affected at dietary concentrations as low as 200 μg kg^{-1}. Histological changes have been observed at the same concentration, but effects on the immune system have not been investigated yet. The fact that Afla and FUM share a common target suggests a high likelihood of a synergistic interaction between these two mycotoxins in the modulation of the sphingolipid metabolism. As this process is vital for rapidly dividing cells such as hemocytes, it would be of interest to investigate the effects of these mycotoxins on different immunologic districts of shrimp.

The effects of DON on the immune system of shrimp have been investigated in several studies, with contradictory results. A study conducted in 2005 on the effects of dietary DON and OTA found no significant effect of 1,000 μg kg^{-1} DON on the hemocyte count, but saw a marked reduction in the activity of the proPO in shrimp that received 1,000 μg kg^{-1} of OTA (Supamattaya *et al.*, 2005). A more recent study conducted in 2018 revealed that DON concentrations up to 1,000 μg kg^{-1} have negative effects on intestinal cell proliferation and the hepatopancreas (Shiwei *et al.*, 2018). The size of hemocytes was reduced, and the number and diameter of hepatopancreatic tubules was reduced as well, thus indicating an immunosuppressive effect of this mycotoxin. The same study observed that immune factors such as proPO, superoxide dismutase and expression of toll-like receptor were increased by dietary DON exposure, indicating a general overstimulation of the immune system in response to DON (Shiwei *et al.*, 2018).

Bundit *et al.* (2006) investigated the effects of dietary ZEN and T-2 on the shrimp immune system. The authors reported that concentrations of up to 2,000 μg kg^{-1} T-2 caused severe necrosis and degeneration of hepatopancreatic tubules, HPT and lymphoid organs. The same effects have been observed for 1,000 μg kg^{-1} ZEN, thus suggesting the possibility of synergistic interactions between these two mycotoxins. Importantly, similar concentrations can be found in feeding crops, even at the inclusion rates currently used in aquafeed.

Based on experience in terrestrial animals and the studies reported here and in Section 2.1, we could conclude that mycotoxins might increase the susceptibility to diseases by weakening the immune system and by interfering with nutrient utilization and thus with the ability of the animal to recover.

More studies investigating the effects of mycotoxins on crustaceans and shrimp are needed to clarify whether these compounds play a determinant role in the modulation of the immune system. The evidence available to date suggests that this might be the case. Given the extreme sensitivity of the shrimp industry to the economic implications that diseases bring along, and given the recent increase in the inclusion levels of plant-based protein sources in shrimp feed, we believe that it is important to pay more attention to the effects of mycotoxin on the immune system, their mode of action and toxicological interactions. Mycotoxins are antinutrients and their presence in feed might jeopardize all the efforts placed into biosecurity and disease control.

03

Mycotoxins in aquaculture

Rui A. GONÇALVES

3. Mycotoxins in aquaculture

Rui A. GONÇALVES

Research characterizing the adverse effects of mycotoxins on the performance and health of animals has in large part focused on terrestrial livestock species (D'Mello and Macdonald, 1997; Pestka, 2007; Rotter *et al.*, 1996). However, since an aflatoxicosis outbreak in trout in the 1960s, research has also been carried out on the effects of mycotoxins in aquaculture species (Wolf and Jackson, 1963). This line of research became even more important in recent years as increasing costs of fish meal necessitate exploring more economical protein sources, such as plant protein or other commercially available plant by-products, which are prone to mycotoxin contamination (e.g. dried distiller's grains with solubles [DDGS]) (Anater *et al.*, 2016; Gonçalves *et al.*, 2018d, 2017; Hooft *et al.*, 2011; Wang *et al.*, 2016). The awareness of mycotoxin-related issues in the industry has grown as feed manufacturers and producers recognize the presence of mycotoxins in feed and observe their impact on production.

3.1 Mycotoxicoses

Mycotoxicoses are diseases in animals or human caused by exposure to mycotoxins, either by ingestion (in feed or water) or contact with/absorption by the skin.

The effects of mycotoxins in fish and shrimps are diverse, varying from immunosuppression to death, in severe cases, depending on toxin-related (type of mycotoxin consumed, level and duration of intake), animal-related (animal species, sex, age, general health, immune status, nutritional standing) and environmental (farm management, biosecurity, hygiene, temperature) factors (Figure 3.1). Therefore, it is often difficult to trace observed problems back to mycotoxins.

Many scientific publications have reported the effects of mycotoxins in fish or shrimp at different contamination levels, enabling a better understanding of mycotoxin-related ailments (Tables 3.1 and 3.2). However, there are still only few validated clinical symptoms of mycotoxin exposure in fish and shrimps. The majority of the described effects of mycotoxins in fish and shrimp (see review from Anater *et al.*, 2016), are general symptoms and could be attributed to diverse pathologies or challenges, for example, anti-nutritional factors or lectins in the diet (Hart *et al.*, 2010). Two notable exceptions are aflatoxicosis (yellowing of the body surface [Deng *et al.*, 2010]) and increase of the sphinganine to sphingosine ratio due to ingestion of fumonisins (FUM) (Tuan *et al.*, 2003). The most frequently reported clinical manifestations of mycotoxin ingestion are a reduction in growth performance, alteration of hematological (erythrocyte/leucocyte count) or biochemical (alanine aminotransferase [ALT], aspartate transaminase [AST] or alkaline phosphatase [ALP]) blood parameters, liver alterations or the suppression of immune parameters.

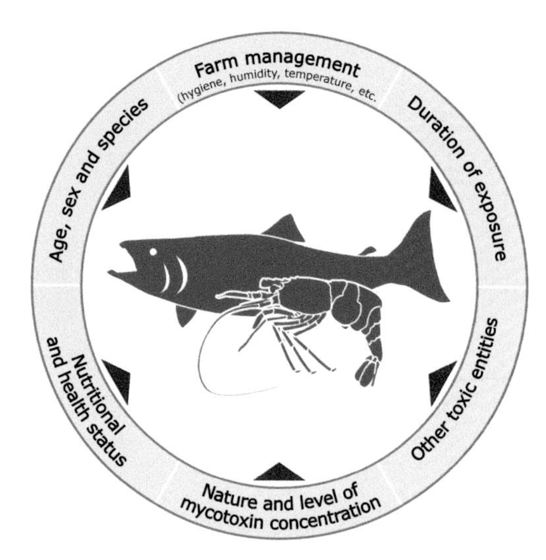

Figure 3.1 *– Interacting factors influencing the effects of mycotoxins in fish and shrimps.*

3.2 Effects of mycotoxins in fish

3.2.1 Aflatoxins

The toxicity of aflatoxins (Afla), mainly of aflatoxin B_1 (Afla B_1), has been the subject of a large number of studies (Table 3.1) in several farmed fish species (Dirican, 2015; Santacroce *et al.*, 2007), including rainbow trout (Arana *et al.*, 2013; Bailey *et al.*, 1994; Carlson *et al.*, 2001; Halver, 1969; Hendricks, 1994; Ngethe *et al.*, 1992, 1993; Ottinger and Kaattari, 1998, 2000; Takahashi *et al.*, 1995), channel catfish (Gallagher and Eaton, 1995; Jantrarotai and Lovell, 1990; Manning *et al.*, 2005a; Plumb *et al.*, 1986), Nile tilapia (Chávez-Sánchez *et al.*, 1994; Deng *et al.*, 2010; El-Banna *et al.*, 1992; Hassan *et al.*, 2010; Oliveira *et al.*, 2013; Tuan *et al.*, 2002; Zychowski *et al.*, 2013b), rohu (Madhusudhanan *et al.*, 2004; Ruby *et al.*, 2013; Sahoo and Mukherjee, 2001a, 2001b), European seabass (El-Sayed and Khalil, 2009), gibel carp (Huang *et al.*, 2014), red drum (Zychowski *et al.*, 2013a), hybrid sturgeon (Rajeev Raghavan *et al.*, 2011) and beluga (Sepahdari *et al.*, 2010). Aflatoxins are especially prevalent in subtropical and tropical areas. They contaminate mainly feedstuffs with high starch and lipid content, such as cottonseed, corn, peanut, wheat and soybean (Ostrowski-Meissner, 1984). Aflatoxin B_1 is known as the most potent carcinogen among Afla, classified as a group I carcinogen by the International Agency for Research on Cancer (IARC, 1993) and is highly hepatocarcinogenic (Busby and Wogan, 1984; Miller and Trenholm, 1994; Sharma and Salunkhe, 1991; Wang *et al.*, 1998, 2008). The biological effects of Afla B_1 in aquatic species are thought to depend on the species, the age of the animal and the concentration of Afla B_1 in feed (Hendricks, 1994). Seabass, the most important aquaculture species in Europe, was reported to be very sensitive to Afla (LC_{50} = 180 μg kg^{-1} body weight [BW]) (El-Sayed and Khalil, 2009). These results were obtained by gavage feeding and it is unclear whether they can be extrapolated to dietary Afla exposure, but they suggest that this species is very sensitive to Afla contaminated feed. In addition, Centoducati and co-authors (2010) concluded that gilthead seabream (*Sparus aurata*) hepatocytes are highly sensitive to Afla B_1 after exposure to $5 \times 10^3 - 2 \times 10^{-5}$ ng Afla B_1 ml^{-1} for 24–72 h. Beluga (*Huso huso*) is another

Table 3.1 *– Overview of literature on effects of aflatoxins, deoxynivalenol, fumonisins, ochratoxins and zearalenone in aquaculture fish species*

Species[a]	**Tested mycotoxin dosages**[b, c]	**Reference**	**Observed signs/symptoms for the tested mycotoxin dosages**[b]	**Experimental design details**
Aflatoxins studies				
Beluga *(Huso huso)*	0; 25; 50; 75[1] and 100[2] μg Afla B_1 kg^{-1} diet	Sepahdari *et al.* (2010)	• Liver lesions[1,2] with fat deposition, hepatocyte degeneration and necrosis[1,2], after 60 days • Decrease in performance (↑FCR[1,2] and ↓WG[1,2])	• Initial weight: 120 ± 10 g; 3 months study
Channel catfish *(Ictalurus punctatus)*	0; 100; 464; 2,154 and 10,000[1] μg Afla B_1 kg^{-1} diet	Jantrarotai and Lovell (1990)	• Tissues lesions in liver[1], head kidney[1], hematopoietic tissue[1], intestines[1], stomach, gastric glands[1] • Decrease in growth rate[1] • Altered hematological parameters[1] (↓Hct, ↓Hb, ↓Ery, ↑HTa)	• Initial weight: 7.5 g; 10 weeks study • [1]Necrotic stomach gastric glands
European seabass *(Dicentrarchus labrax L.)*	[#1]Oral 96 h LC_{50} > 50; 100; 150; 200; 250; 300; 350 and 400 μg kg^{-1} diet [#2]42 days exposure to 10% of oral 96h LC_{50} = 180 μg kg^{-1} diet	El-Sayed and Khalil (2009)	• Tissues lesions in liver[1,2], kidney[1,2], skin[1,2], gills [1,2] and gallblader[1,2] • Altered hematological parameters[2] (↑ALT, ↑AST, ↑ALP, ↓TP, ↓alb, ↓glob) • Accumulation of AF in [2]muscle = 5 μg Afla B_1 Kg^{-1} • [#1,2] Clinical signs: sluggish movement, loss of equilibrium, rapid opercular movement, and hemorrhages of the dorsal skin surface. [#2]Yellowish discoloration, pale discoloration of the gills, liver and kidney. Severe distension of the gall bladder.	• Initial weight: 40 ± 2 g • [#1]96h LC_{50} = 0.18 mg/kg bwt • [#2]Intake: 18 mg/kg bwt Afla B_1 (10% of oral 96h)
Gibel carp *(Carassius gibelio)*	3.3[1]; 22.3[2] and 1,646.5[3] μg Afla B_1 kg^{-1} diet	Huang *et al.* (2014)	• Decreased gonadsomatic index[2,3], decreased absolute brood amount[2,3], decreased relative fecundity[2,3] and decreased oocyte diameter[2,3] • Accumulation of Afla in: ovary[1] = 1.20 μg Afla B_1 Kg^{-1}, ovary[2]= 4.12 μg Afla B_1 Kg^{-1}, ovary[3]= 5.32 μg Afla B_1 Kg^{-1}, muscle[2] = 3.11 μg Afla B_1 Kg^{-1}, muscle[3] = 4.0 μg Afla B_1 Kg^{-1}	• Initial weight: 57.6 ± 0.1 g • 24 weeks study

***Table 3.1** – Contd.*

Gilthead sea bream *(Sparus aurata)*	From 5×10^{3} ng Afla B_1 ml^{-1} to 2×10^{-5} ng Afla B_1 ml^{-1}	Centoducati *et al.* (2010)	• Cytotoxic effect on primary monolayer cultures of hepatocytes from *S. aurata* juveniles • Exhibited dose- and time-dependent cytotoxic effect, the IC_{50} being inversely related to the exposure time (MTT-IC_{50}-24h, $5x10^{3}$ ng/ml; MTT-IC_{50}-48h, $6x10^{2}$ ng/ml; MTT-IC_{50}-72h, 60 ng/ml)	• Exposure = 4, 48, 72 hours
Hybrid sturgeon (*Acipenser ruthenus x A. baeri*)	0; 1; 5; 10; 20^{1}; 40^{2} and 80^{3} µg Afla B_1 kg^{-1} diet	Rajeev Raghavan *et al.* (2011)	• $Liver^{2,3}$ hypertrophy and hyperchromasia of nuclei and cytoplasmic vacuoles, presence of inflammatory cells, focal hepatocyte necrosis and extensive biliary hyperplasia. • Accumulation of Afla in liver = 142.80^{2} and 115.60^{3} ng g^{-1}	• Initial weight: 10.53 ± 0.17 g • 35 days study
Hybrid tilapia *(Oreochromis niloticus × O. aureus)*	19; 85^{0}; 245^{1}; 638^{2}; 793^{3} and $1,641^{4}$ µg Afla B_1 kg^{-1} diet	Deng *et al.* (2010)	• Lesions found in $liver^{t;\ 2\text{-}4}$, yellow $skin^{3,4;\ >t1}$ • Decrease in growth $performance^{t;\ 1\text{-}4}$ (↓WG, ↓FI, ↓FER) • Altered hematological $parameters^{t;\ 2\text{-}4}$ (↓TP, ↓alb) • Accumulation of Afla in $tissues^{t1+t;\ 0\text{-}4}$ $Liver^{t1}$=10^{0}, 16^{1}, 21^{2}, 24^{3} and 24^{4} µg Afla B_1 kg^{-1} liver $Liver^{t}$=30^{0}, 33^{1}, 47^{2}, 44^{3} and 43^{4} µg Afla B_1 kg^{-1} liver	• Initial weight: 20 g; • 20^{t} weeks study (sampling at week 5^{t1}) • Afla from moldy peanut meal
Nile tilapia *(Oreochromis niloticus)*	0; 940; $1,880^{\#1}$; 375; 752^{1}; $1,500^{2}$; $3,000^{3}$ µg Afla B_1 kg^{-1} diet	Chávez-Sánchez *et al.* (1994)	• Lesions found in $liver^{2,3}$, $kidney^{2,3}$and $gills^{1\text{-}3}$. Fatty liver and characteristic neoplastic changes such as nuclear and cellular hypertrophy, nuclear atrophy, increase in number of nucleoli, cellular infiltration, hyperemia, cellular basophilia and necrosis. • Decrease in $performance^{1\text{-}3}$ (↓FI, ↓WG, ↓SGR)	• Initial weight: 0.5 g • 25 days study with extra 50 days recover period • Sampling at days 15, 26, 54 and 75 $^{\#1}$ - Afla value seems to be error in the manuscript

Nile tilapia *(Oreochromis niloticus)*	0; 1.5[1] µg Afla B_1 kg^{-1} diet	Hassan *et al.* (2010)	• Lesions found in liver[1] • Altered hematological parameters[1] (↓TP, ↓alb, ↓glo, ↓test)	• Initial weight: 90 ± 10 g; 3 weeks study • Afla effects are reported only for 1.5[1] µg kg^{-1}
Nile tilapia *(Oreochromis niloticus)*	350; 757; 1,177[1] µg Afla B_1 kg^{-1} diet	Oliveira *et al.* (2013)	• Decrease in performance[#2] (↓FCR[1] and ↓TL[1])	• Initial weight: 1.55 ± 0.005g[#1]; 30 days study • Challenge test[#2] with *Aeromonas hydrophila* • Authors suggest a synergism between Afla B_1 and *Aeromonas hydrophila.*
Nile tilapia *(Oreochromis niloticus)*	0; 250[1]; 2,500[2]; 10,000[3] or 100,000[4] µg Afla B_1 kg^{-1} diet	Tuan *et al.* (2002)	• Tissue lesions were found in liver[3,4], no lesions[1-4] were found in spleen, stomach, intestines or kidney • Decrease in weight gain[2-4] and in haematocrit[2-4]	• Initial weight: 2.7g • 8 weeks study
Nile tilapia (Oreochromis niloticus)	0; 1,500[1] or 3,000[2] µg Afla B_1 kg^{-1} diet	Zychowski *et al.* (2003b)	• Tissue lesions were found in liver[1,2] • Decrease in performance[1,2] (↓WG, ↓FI, ↓HIS)	• Initial weight: 2 ± 0.1 g; 10 weeks study
Rainbow trout *(Oncorhynchus mykiss)*	0 and 80,000[1] µg Afla B_1 kg^{-1} diet	Arana *et al.* (2013)	• Tissue lesions were found in liver[1]	• Initial weight: 18 g; 12 months study • 6 animals sampled monthly
Rainbow trout *(Oncorhynchus mykiss)*	[#1]0.01[1]; 0.025[2]; 0.05[3]; 0.1[4]; 0.25[5]; 0.5[6] µg Afla B_1 kg^{-1} diet [#2]4[1]; 8[2]; 16[3]; 32[4]; 64[5] µg Afla B_1 kg^{-1} diet	Bailey *et al.* (1994)	• Increase in hepatic tumor incidence (Hti), [#1]↑Hti[1-6] and [#2]↑Hti[1-5]	• [#1]21 day old embryos exposed by bath to Afla for 1 h • [#2] Dietary supplementation of Afla B_1 for 9 months
Red Drum *(Sciaenops ocellatus)*	0; 100[1]; 250[2]; 500[3]; 1,000[4]; 2,000[5]; 3,000[6] or 5,000[7] µg Afla B_1 kg^{-1} diet	Zychowski *et al.* (2013a)	• Tissue lesions were found in liver [3-7] • Decrease in feed efficiency[1,2, 5-7] • Altered hematological parameters (↓Lyz[1-7], ↓Try[1,3-7], ↓HSI[5,7])	• Initial weight: 2.1 ± 0.1 g, 7 weeks study

***Table 3.1** – Contd.*

Rohu *(Labeo rohita)*	0; 100^{1} µg Afla B_1/100 g and 100 µg Afla B_1/100 g	Madhusudhana *et al.* (2004)	• Tissue lesions were found in liver^{1}, kidney^{1} and brain^{1} • Increase in conjugated diene formation and LPO not only in liver but also in kidney and brain	• Initial weight: 175–250 g, 7 weeks study • Intravenous injection and oral administration
Rohu *(Labeo rohita)*	Control, 10% Moldy Feed^{1} MF, 50% MF^{2} and 100% MF^{3}	Ruby *et al.* (2013)	• Decrease in performance (↓FBW$^{1-3}$ and ↓SGR$^{1-3}$)	• Initial weight: 30–50 g • Afla produced from moldy feed • Afla contamination values were not determined. Authors performed the experiment with inclusion of moldy feed (MF) • The authors did not treated data statistically
Rohu *(Labeo rohita)*	0; 1,250^{1}; 2,500^{2} and 5,000^{3} µg Afla B_1 kg^{-1} diet	Sahoo and Mukherjee (2001a)	• Altered hematological parameters (↓NTB^{2,3}, ↓TP$^{1-3}$, ↓glob$^{1-3}$, ↓A:G$^{1-3}$)	• Initial weight: 30–50 g, 90 days study • Intravenous injection of Afla B_1 • Bacterial agglutination titer with *Edwardsiella tarda*
Rohu *(Labeo rohita)*	0; 1,250^{1} and 5,000^{2} µg kg^{-1} Afla B_1 BW	Sahoo and Mukherjee (2001b)	• Altered hematological parameters (↓NTB^{2}, ↓TP^{1,2}, ↓Alb^{2}, ↓glob^{2}, ↑A:G$^{1-3}$)	• Initial weight: 30–50 g, 90 days study • Intravenously injection of Afla B_1 and observed in the subsequent 90 days; • Bacterial agglutination titer with *Edwardsiella tarda*
Tra catfish *(Pangasius hypophthalmus)*	0; 62^{1}; 104^{2}; 237^{3}; 468^{3}; and 945^{4}µg Afla B_1 kg^{-1} diet	Gonçalves *et al.* (2018a)	• Decrease in performance (↓WG$^{T1-T3; 1-5}$, ↑FCR$^{T1-T3; 3-5}$, ↓SGR$^{T1-T3; 1-5}$) • Altered hematological parameters (↓Ery$^{T2-T3; 1-3}$, ↑Leu$^{T2-T3; 1-3}$, ↑HSI$^{T2-T3; 3}$, ↑ASI$^{T2-T3; 3}$, ↑AST$^{T2-T3; 3}$, ↑ALT$^{T2-T3; 2,3}$)	• Initial weight: 8.0 ± 0.2 g; 12 weeks study; sampling after 4^{T1}, 8^{T2} and 12^{T3} weeks • Treatments ^{1,2} and ^{3} challenged with *Edwardsiella ictaluri* by bath (additional information about tested products in supplementary notes)

Yellow catfish *(Pelteobagrus fulvidraco)*	0; 200[1]; 500[2]; and 1,000[3] µg Afla B_1 kg^{-1} diet	Wang *et al.*, (2016)	• Altered hematological parameters (↑GPT[3], ↑GOT[3], ↑A:G[3], ↓Lyz[3], ↓RBC[1-3], ↓TP[3] and ↓bact.aw[3]) • Decrease in performance (↑FCR[2,3], ↓FBW[3], ↓WG[3])	• Initial weight: 2.02–0.10 g, 12 weeks study
Deoxynivalenol studies				
Atlantic salmon *(Salmo salar)*	Control; 4x [DON] 800–3,700[1] µg DON kg^{-1} diet	Döll *et al.* (2010)	• Decrease in performance (FI[1], ↓SGR[1], ↑FCR[1])	• Initial weight: 405 ± 31 g, 15 weeks study
Channel catfish *(Ictalurus punctatus)*	0; 3,300[1]; 5,500[2]; 7,700[3]; 8,800[4] µg DON kg^{-1} diet	Manning *et al.* (2014)	• Decrease in performance (↓FCR[4])	• Initial weight: 5.87 ± 0.22 g, 7 weeks study • Naturally contaminated corn (DON) mixed with clean corn • Challenge[C] by bath with *Edwardsiella ictaluri* (21 days infection)
Common carp *(Cyprinus carpio)*	352[1]; 619[2] and 953[3] µg DON kg^{-1} diet	Pietsch *et al.* (2014)	• Altered hematological parameters (↓EryL[1], ↓Ery.N.W[2], ↑SOD[1, RP], ↑CAT[1, RP]) • Accumulation of DON in muscle samples [ng g^{-1} dry weight] [1]= 0.6; [1, RP]= 1.4; [1]= 1.3; [1, RP]= 0.7; [1]= 1.2; [1, RP]= 0.0	• Raised from eggs (average initial weight 36 g), 4 weeks study • Leaching test – 50% of DON leaches in first 2h • Additional 2 weeks of feeding uncontaminated diet – recovery period[RP]
Rainbow trout *(Oncorhynchus mykiss)*	[#1]0; 1,166[1] and 2,745[2] µg DON kg^{-1} diet [#2]0; 367[1] µg DON kg^{-1} diet	Gonçalves *et al.* (2018)	• Tissue lesions were found in liver [#1:2] • Decrease in performance ([#1]↓FBW [2], [#1]↓SGR [2], [#1]↓SGR [2], [#1]↓FI [2], [#1]↓TGC [2], [#2]↓FBW [Tf, 1]) • Altered hematological parameters ([#1]↑ALT [2], [#1]↑AST [2])	• [#1]Experiment 1: Initial weight: 14.10 ± 0.05 g; 50 days study • [#1] Trout challenged[C] by bath with *Yersinia ruckeri* • [#1] Some trout fed[2] showed abnormal body conformations and protruding anal papilla • [#2]Experiment 2: Initial weight: 89 ± 8 g; 168 days study; sampling on days 37, 62, 92, 125 and 168[Tf]

Table 3.1 – *Contd.*

Rainbow trout *(Oncorhynchus mykiss)*	37.42^{1}; 4,714^{2}; 11,412^{3} µg DON kg^{-1} diet	Gonçalves *et al.* (2018)	• Altered gene expression (↓*igf1* $^{T2, 2,3}$, ↓*igf2* $^{T2, 2,3}$, ↑*npy* $^{T2, 2,3}$, ↑*adcyap1a* $^{T2, 2,3}$, ↑*try3* $^{T2, 2,3}$) • Decrease in performance (↓FBW $^{T1,T2, 2,3}$, ↓SGR $^{T1,T2, 2,3}$, ↑FCR $^{T1,T2, 3}$, ↓FI $^{T1,T2, 2,3}$, ↓PER ^{T1,2,3}, ↓PER ^{T2,3}) • No clinical signs (except anorexia at the higher DON dosage) were observed	• Initial weight: 2.52 ± 0.03 g; 60 days study; sampling at 29^{T1} and 60 ^{T2} days. • Experimental diets were contaminated with ZEN (78.63$^{1-3}$) and FB_1 (67.73$^{1-3}$) in addition to DON. • DON is metabolized to DON-3-sulfate
Rainbow trout *(Oncorhynchus mykiss)*	300^{1}; 800^{2}; 1,400^{3}; 2,000^{4}; 2,600^{5} µg DON kg^{-1} diet	Hooft *et al.* (2011)	• Tissue lesions were found in liver$^{3-5}$ • Decrease in performance (↓FE; ↓FI; ↓TGC; ↓WG; ↓RN; ↓RE; ↓NRE; ↓ERE)	• Initial weight: 24.3 g, 8 weeks study (samplings every 28 days)
Rainbow trout *(Oncorhynchus mykiss)*	300^{1}; 1,000^{2}; 1,500^{3} and 2,000^{4} µg DON kg^{-1} diet	Hooft *et al.* (2017)	• Decrease in performance (↓FI; ↓TGC; ↓WG; ↓RN; ↓RE; ↓NRE; ↓ERE)	• Initial weight: 1.8 g, 12 weeks study (samplings every 28 days) • No differences in feed efficiency
Rainbow trout *(Salmo gairdneri)*	0^{0}; 19,400^{1}; 40,400^{2}; 55,300^{3}; 84,300^{4}; 109,600^{5} µg DON kg^{-1} diet	Woodward *et al.* (1983)	• Decrease in performance (↓FI$^{1-5;t1}$, ↓WG$^{1-5;t1}$, ↓FCR$^{1-5;t1}$) • Feed refusal$^{1-5}$; FI got normal when diets were changed	• No fish weight available; fry stage; 8 weeks study • Fish fed contaminated levels for 4^{t1} weeks. After 4 weeks diets were changed to 1,(940^{0}), 0^{1}, 4,(040^{2}), 0^{3}, 8,(430^{4}), 0^{5}, respectively, to test recovery. • Natural contamination from mixture of ground husks, cobs and kernels, and contained approximately 4 µg zearalenone g^{-1}, a trace of 7-deoxyvomitoxin, but un-detectable levels of numerous other mycotoxins including nivalenol, 3-acetyldeoxynivalenol, fusarenone-X, diacetoxyscirpenol, T-2 toxin, neosolaniol and HT-2 toxin

Fumonisins studies				
Atlantic salmon *(Salmon salar)*	1,000; 5,000; 10,000 or 20,000 μg FUM kg^{-1} diet	Garcia (2013)	• No alteration of measured parameters (histology, FI, FW, WG, FCR, SGR)	• Initial weight: 31.8 ± 6.4 g, 10 weeks study • mycotoxin levels in feeds were not confirmed/analyzed or it was not reported
Channel catfish *(Ictalurus punctatus)*	300; 20,000[1]; 80,000[2]; 320,000[3]; 720,000[4] μg FUM kg^{-1} diet	Lumlertdacha and Lovell (1995)	• Tissue lesions were found in liver[1-4;wk14] • Decrease in performance (↓WG[3,4;W2-W14], ↓WG[2;W4-W14]) • Altered hematological parameters (↓Hct[3,4;W14], ↓RBC[3,4;W14], ↓WBC[3,4;W14]) • Increased mortality[3;W12,W14] and [4;W14] • Clear age-related effect	• Initial weight: 31 g, 14 weeks study (sampling every 2 weeks; week 2– week 14) • Naturally contaminated corn containing 1,600 mg FB_1 kg^{-1}
Common Carp *(Cyprinus carpio)*	0; 10,000[1]; 100,000[2] μg FUM kg^{-1} diet	Petrinec *et al.* (2004)	• Tissue lesions were found in liver[1,2], gallbladder[1,2], head kidney[1,2], kidney[1,2] and brain[1,2]	• Initial weight: 15.87 g, 4 weeks study • Petechial hemorrhages and edema in kidney
Common Carp *(Cyprinus carpio)*	0; 500[1]; 5,000[2] μg FUM kg^{-1} diet	Pepeljnjak *et al.* (2003)	• Tissue alteration: dilation of sinudoids[2] • Decrease in performance (↓WG[1,2]) • Altered physiological parameters (↑RBCs[1,2], ↓MCV[2], ↑RBCs[1,2], ↑Bir[1,2], ↑AST[1,2], ↑ALT[1,2], ↑TP[1,2], ↑Creat[1,2])	• Initial weight: 120–140 g, 42 days study • Diets contaminated with *F. moniliforme* corn culture material • No effects on: WBCs, Hb, Hct, MCH and MCHC
Nile tilapia *(Oreochromis niloticus)*	0; 10,000[1]; 40,000[2]; 70,000[3]; 150,000[4] μg FUM kg^{-1} diet	Tuan *et al.* (2003)	• Decrease in performance (↓WG[2,4;wk4-wk8], ↑FCR[2,4;wk8]) • Altered physiological parameters (↓Hct[4;wk8], ↑Sa/So[4;wk8]) • No histological changes in liver, spleen, stomach, intestines and kidney	• Initial weight: 2.7 g, 8 weeks study (sampling every 2 weeks; week 2 – week 8) • Diets contaminated by *F. moniliforme* corn culture material
Nile tilapia *(Oreochromis niloticus)*	0; 20,000[1]; 40,000[2] and 60,000[3] μg kg^{-1} $FB_1 + FB_2$	Claudino-Silva *et al.* (2018)	• Altered physiological parameters (↓[L]IGF-1[1-3], ↓[L]GHR[1-3]) • Decrease in performance (↓WG[1-3])	• Initial weight: 2.64 ± 0.06 g • 30 days study, with sampling after 15[T1] and 30 [T2] days

Table 3.1 – *Contd.*

Rainbow trout *(Oncorhynchus mykiss)*	0; 600[1], 20,000[2]; 63,000[3] μg FUM kg^{-1} diet	Meredith *et al.* (1998) and Riley *et al.* (2001)	• Altered physiological parameters (↑Sa/So) • Tissue lesions were found in liver[2,3], kidney[2-3], serum[2-3]	• Initial weight: 15–25 g, 1 week study
Ochratoxins studies				
Channel catfish *(Ictalurus punctatus)*	0; 500[1]; 1,000[2]; 2,000[3]; 4,000[4]; 8,000[5] μg OTA kg^{-1} diet	Manning *et al.* (2003)	• Tissue lesions were found in liver and kidney. • Decrease in performance (↓WG[2-5,WK2,8]; ↓WG[3-5,WK4,6]; ↓FCR[4,5]) • Altered hematological parameters (↓Htc[5]; ↑WBCs) • Increased mortality[5]	• Initial weight: 6.1 g, 8 weeks study (sampling every 2 weeks; week 2 – week 8) • Diets contaminated with *Aspergillus ochraceus* culture material
Channel catfish *(Ictalurus punctatus)*	0; 2,000[1] and 4,000[2] μg OTA kg^{-1} diet	Manning *et al.* (2005b)	• Decrease in performance (↓WG[1,2]) • Increased mortality after challenge[C; 1,2]	• Initial weight: 6.4 g, 6 weeks feeding study • Challenge[C] (by bath) with *Edwardsiella ictaluri* for 21 days (estimated 2.25 x 10^6 colony forming units per ml of water)
Common carp *(Cyprinus carpio)*	Control[C]=15 + 22 **Afla B_1/OTA** [P0.2%]=12 + 14 **Afla B_1/OTA** [P0.4%]=9 + 6 **Afla B_1/OTA** [B0.15%] =21 + 13 **Afla B_1/OTA** [B0.25%] =21 + 15 **Afla B_1/OTA** μg kg^{-1}	Agouz and Anwer (2011)	• Decrease in performance (↓FW[C], ↓TWG[C], ↓SGR[C], ↑FI[C], ↑FCR[C], ↓PER[C]) • Altered hematological parameters (↓Htc[C], ↑WBCs[C],↑RBCs[C], ↓Hb[C])	• Initial weight: 15 g, 8 weeks study (sampling every 2 weeks; week 2 – week 8) • Tested a probiotic composed of *Bacillus subtilis*, proteolytic, lipolytic amylolytic and cell separating enzymes, germanium and organic selenium (inclusion levels 0.2%[P0.2%] and 0.4% [P0.4%]) • Tested a commercial smectite clay composed of sodium/ calcium aluminosilicate (HSCAS; inclusion levels 0.15%[B0.15%] and 0.25% [B0.25%]) • Control diets showed the highest concentrations of AF and OTA; evaluation refers only to control treatment.

European seabass (*Dicentrarchus labrax*)	0; 50; 100; 150; 200; 250; 300; 350 and 400 µg OTA kg^{-1} diet	El-Sayed *et al.* (2009)	• Estimated lethal concentrations (µg kg^{-1}): LC_1=146.2; LC_5=176.3; LC_{10}=195.7; LC_{15}=208.3; LC_{50}=277.0; LC_{85}=368.7; LC_{90}=393.2; LC_{95}=434.4	• Initial weight: 40 ± 2 g, sampling after 24, 48, 72 and 96 h • OTA contaminated feed was administered to individual animals once daily by oral gavage
Rainbow trout (*Salmo gairdneri*)	**OTA**: 2,000[1]; 3,000[2]; 4,000[3]; 5,000[4]; 6,000[5] and 8,000[6] µg OTA per kg^{-1} body weight **OTB**: 16,700[7]; 33,300[8] and 66,700[9] µg OTB per kg^{-1} body weight OTα: 7,000[10]; 14,000[11] and 28,000[12] µg OTα per kg^{-1} body weight **OTβ**: 6,700[13]; 13,300[14] and 26,700[15] µg OTβ per kg^{-1} body weight	Doster *et al.* (1972)	• Severe degeneration and necrosis of kidney[1-6] and liver[1-6]; pale kidney[1-6], light swollen livers[1-6] and mortality was observed[1-6] • Estimated lethal concentrations: OTA: LC_{50} = 4,670	• Initial weight: 30 g, 10 days study • Diets contaminated with *Aspergillus ochraceus* culture material from wheat • single intra-peritoneal dose of each of the four ochratoxins
Zeralenone studies				
Common Carp (*Cyprinus carpio*)	0; 332[1]; 621[2] and 797[3] µg ZEN kg^{-1} diet	Pietsch *et al.* (2015)	• Altered hematological parameters (↓Mono[2,3]; ↓gran[2,3]; ↓MNery[1-3]) • Performance parameters were not modified (FW; WG; SGR; FCR; CF) • Accumulation of ZEN and its metabolites in muscle[1-3] : ZEN=0.13[1]/0.22[2]/0.15[3] ng g^{-1} dry weight and α-ZEN = 0.11[1]/0.16[2]/0.05[3] ng g^{-1} dry weight	• Raised from egg, used with 12–16 cm in length • 4 weeks study • α-ZEN was not detectable after recovery period (2 weeks) and ZEN was detected at 0.03 ng g^{-1} dry weight for all treatments

Table 3.1 *– Contd.*

Rainbow trout *(Oncorhynchus mykiss)*	10,000 µg ZEN kg^{-1} of body mass	Woźny *et al.* (2012)	• Altered hematological parameters (↑Bct; [Fe]:↓$L^{24-168h}$; ↓$O^{24-168h}$)	• Initial weight: 53.3 ± 5.3 g, all mature females; • Injected intraperitoneally with ZEN • Sampling after 24 h, 72 h, 168 h
Zebrafish *(Danio rerio)*	0; 100[1]; 320[2]; 1,000[3] and 3,200[4] ng l^{-1}	Schwartz *et al.* (2010)	• Several reproductive parameters were altered (↑$Vtg^{3,4}$; ↓$SF^{PE,E}$; ↓$CS^{1,2,4;PE,E}$; ↓$Fecund^{2-4;PE,E}$) • Gonad morphology were not altered.	• Spawning groups (two females and four males) • Pre-exposure[PE] period of 21 days and 21 days exposure[E] during spawning • Spawned eggs were removed from exposure tanks and examined
Zebrafish *(Danio rerio)*	0; 0.1[1]; 0.32[2] and 1[3] µg l^{-1}	Schwartz *et al.* (2013)	• Several reproductive parameters were altered (↑Lg^{H}; ↑LW^{H}; ↑$CF^{Hc;H}$; ↓♂/♀^{M,H}; ↑$RFecund^{H}$; ↑$VTGc^{H}$)	• Whole life cycle (eggs to adult; 224 days) • ZEN solution was supplied by means of a computer-controlled dispenser at a rate of 600, 1,920, and 6,000 µl h^{-1} giving nominal exposure concentrations of 0.1, 0.32 and 1 µg l^{-1}, respectively. • Experimental groups: $Control^{C}$; Low (0.1 µg l^{-1}) during reproduction (F0), juvenile (F1) and reproduction (F1) –L^{L} or Lc^{Lc} (recovery during juvenile phase): medium (0.32 µg l^{-1}; M^{M} or Mc^{Mc}) and high (1 µg l^{-1}; H^{H} or Hc^{Hc}).
Zebrafish *(Danio rerio)*	0.1[1]; 10[2]; 1,000[3] µg l^{-1}	Bakos *et al.* (2013)	• Several reproductive parameters were altered (↑Vtg^{3}; ↑*vtg-1* $mRNA^{2,3}$)	• Adult fish; 21 days study • Exposure by bath • Sexually mature males

Rainbow trout *(Oncorhynchus mykiss)*	1,810 μg ZEN kg^{-1} diet	Woźny *et al.* (2015)	• Hematological parameters were not modified (TP; Alb; AST; ALP; TAG) • Accumulation of ZEN and its metabolites in the intestines (ZEN = 732.2 μg kg^{-1} and α-ZEN = 10.7 μg kg^{-1}); residual ZEN and α-ZEL in liver of all sampled fish	• Initial weight: 250 g, all females; 71 days study • Some animals were identified as males • ZEN was detected (< 5.0 μg·kg^{-1}) in all females' ovaries

Notes: [a] Studies are ordered alphabetically by common species name.

[b] Red superscript numbers and letters link methods (e.g. specific doses tested) to results of the same study (i.e. presented in the same row).

[c] Purified mycotoxin was applied unless stated otherwise.

Mycotoxins: Afla: aflatoxins (the sum of Afla B_1, Afla B_2, AFG_1 and AFG_2); Afla B_1= aflatoxin B_1; Afla B_2= aflatoxin B_2; AFG_1= aflatoxin G_1; AFG_2= aflatoxin G_2; DON = deoxynivalenol; FUM = fumonisins (the sum of FB_1 and FB_2); FB_1= fumonisin B_1; FB_2= fumonisin B_2; OTA= ochratoxin A; ZEN= zearalenone.

Other abbreviations: A:G – Albumin/globulin ratio; *adcyap1a* – growth hormone-releasing hormone/pituitary adenylate cyclase-activating polypeptide (PACAP); Alb – Albumin; ALT – Alanine aminotransferase; ASI – Aspartate aminotransferase; AST – Aspartic aminotransferase; Bact.aw – Bactericidal activity; Bct – Blood clotting time; Bir – Bilirubin; CAT – Catalase; Cf – Condition factor; Creat – creatin; ERE – energy retention efficiency; Ery – Erythrocyte count; Ery.N.W – Erythrocyte Nucleus width; EryL – Erythrocyte length; FBW – Final body weight; FCR – Feed conversion ratio; FI – Feed intake; Glob – Globulin; GOT – Glutamic-oxaloacetic transaminase; GPT – Glutamic-pyruvic transaminase; Hb – hemoglobin; Hct – hematocrit; HIS – hepatosomatic index; HTa – hematopoietic tissue activity; Hti – hepatic tumor incidence; IC – Inhibitory concentration; *IGF1* – insulin-like growth factor 1; *IGF2* – insulin-like growth factor 2; Leu – leucocytes; Lg – Length; LPO – lipid peroxidation; Lw – Liver weight; Lyz – Plasma lysozyme concentration; MCV – mean erythrocyte volume; *NPY* – neuropeptide Y precursor; NRE – nitrogen retention efficiency; NTB – Nitroblue tetrazolium assay; PER – Protein efficiency ratio; RBCs – Total erythrocyte count (red blood cells); RE – recovered energy; RFecund – relative fecundity; RN – retained nitrogen; Sa/So – sphinganine/ sphingosine ratio; SGR – Specific growth rate; SOD – Superoxide dismutase; Test – testosterone; TGC – Thermal growth coefficient; TL – Total length; TP – Total protein; *TRY3* – trypsinogen 3 precursor; Vtg – Vitellogenin; *VTG-1* – vitellogenin 1; VTGc – Vitellogenin concentration; WG – weight gain.

Table 3.2 *– Overview of literature on effects of aflatoxins, deoxynivalenol, fumonisins, ochratoxins and zearalenone in aquaculture shrimp species*

Species[a]	Tested mycotoxin dosages[b,c]	Reference	Observed signs/symptoms for the tested mycotoxin dosages[b]	Experimental design details
Aflatoxins studies				
Black tiger shrimp *(Penaeus monodon)*	0; 5[1]; 10[2] and 20[3] μg kg^{-1} Afla B_1 diet	Bintvihok *et al.* (2003)	• Tissue lesion were found in hepatopancreas [1-3] • Decrease in performance (↓FBW[1-3;t2]) • Altered physiological parameters (↓GOT, ↓GTP) • Increased mortality [2,3;t1; 1-3;t2]	• 3.5 months old, 11 days study (samples at 8 [t1] and 11[t2] days) • Afla B_1 was prepared from moldy corn
Black tiger shrimp *(Penaeus monodon)*	0; 50[1]; 100[2]; 500[3]; 1,000[4]; 2,500[5] μg kg^{-1} Afla B_1 diet	Boonyaratpalin *et al.* (2001)	• Tissue lesions were found in hepatopancreas [3-5] • Decrease in performance (↓FBW[4,5]; ↓WG[5]) • Increased mortality [5]	• Study in juvenile stage = 0.7 g; 8 weeks trial • Reddish discoloration dispersed over body and tail for 2,500 μg kg^{-1} Afla B_1 treatment • A bacterial suspension of *Vibrio harveyi* was injected into the tail muscle
Pacific white shrimp *(Litopenaeus vannamei)*	53,000[1]; 75,000[2]; 106,000[3]; 150,000[4]; 300,000[5] μg kg^{-1} Afla B_1 diet	Wiseman *et al.* (1982)	• Tissue lesions were found in hepatopancreas[1-3], mandibular organ[1-3] and hematopoietic organs[1-3]	• Initial weight: 2.6 ± 0.5 g, 10 days study • Intravenous injection of Afla B_1 • 24 h LD_{50} = 100,500 μg kg^{-1} Afla B_1 (95% confidence interval: 78.3 – 129.0) • 96 h $LD5_{50}$ = 49,500 μg kg^{-1} Afla B_1 (95% confidence interval: 29.8–82.3)
Pacific white shrimp *(Litopenaeus vannamei)*	0; 50[1]; 250[2]; 1,500[3]; 3,000[4]; 15,000[5] μg kg^{-1} Afla B_1 diet	Ostrowski-Meissner *et al.* (1995)	• Tissue lesions were found in hepatopancreas [1-5] and antennal gland[1-5] • Decrease in performance (↓FBW[3-5], ↑FCR[1-5]) • Increased mortality [5]	• Initial weight: 1.61 ± 0.19 g, 21 days study • Survival at 15 ppm = 0%

Pacific white shrimp *(Litopenaeus vannamei)*	15[1]; 20[2]; 60[3]; 300[4]; 400[5]; 900[6] μg kg^{-1} Afla B_1 diet	Ostrowski-Meissner *et al.* (1995)	• Decrease in performance (↓FBW[5,6]; ↑FCR[5,6])	• Initial weight: 1.51 ± 0.05 g, 8 weeks study • Dry matter apparent digestibility coefficients, crude protein and digestible energy was decreased in 900 μg kg^{-1} Afla B_1 treatment
Pacific white shrimp *(Litopenaeus vannamei)*	0; 75[1] μg kg^{-1} AF	Tapia-Salazar *et al.* (2017)	• Decrease in performance (↓GR[1], ↑Nr[1])	• Initial weight: 210 ± 4 mg • Afla B_1 contamination achieved by formulating diets with contaminated corn (7,130 μg kg^{-1}; NuteK S.A)
Pacific white shrimp *(Litopenaeus vannamei)*	0; 15,000 μg kg^{-1} Afla B_1 diet	Zhao *et al.* (2017)	• Tissue lesions were found in hepatopancreas [T3] • Altered physiological parameters (↑SOD [T3], ↑GST [T3], ↑GPx [T3], ↑CAT [T3]) • 1,024 genes were differentially expressed in shrimp fed Afla B_1, being involved in functions such as peroxidase metabolism, signal transduction, transcriptional control, apoptosis, proteolysis, endocytosis, and cell adhesion and cell junctions	• Initial weight: 2.40 ± 0.13 g; 12 days study • Shrimps were sampled at 1[T1], 2 [T2], 8 [T3] and 12 [T4] days
Pacific white shrimp *(Litopenaeus vannamei)*	0; 15,000 μg kg^{-1} Afla B_1 diet	Zhao *et al.* (2018)	• Tissue lesions were found in hepatopancreas [T3] • Increased mortality [T2, T3] • Several genes were differentially expressed in shrimp fed Afla B_1, being involved in metabolic functions, including the metabolism of pyrimidine, purine, mannose, arginine, proline, glycine, serine, galactose, sphingolipids, valine, leucine and isoleucine, and fatty acids	• Initial weight: 2.40 ± 0.13 g; 12 days study • Shrimps were sampled at 1[T1], 6 [T2] and 12 [T3] days
Pacific white shrimp *(Litopenaeus vannamei)*	0 and 500[1] μg kg^{-1} Afla B_1 x 100 and 200 mg kg^{-1} Zn-CM	Yu *et al.* (2018)	• Tissue lesions were found in hepatopancreas [1] and hepatopancreatic sloughing [1] • Decrease in performance (↓WG[1]) • Altered physiological parameters (↑MDA[1], ↑GSH[1], ↑ALT[1])	• Initial weight: 1.08 g; 8 weeks study • During acclimation (2 weeks) shrimp were fed a commercial diet with 8.3 μg kg^{-1} Afla B_1

Table 3.2 *– Contd.*

Deoxynivalenol studies				
Black tiger shrimp *(Penaeus monodon)*	500^{1}; $1{,}000^{2}$ and $2{,}000^{3}$ µg DON kg^{-1} diet	Supamattaya *et al.* (2005)	• Decrease in performance (↓FR^{1-3}, ↑$Gr^{1,2}$, ↓SGR^{1-3}) • Altered physiological parameters (↓ALP^{1-3}, ↓$SGOT^{2,3}$, ↓$SGPT^{2,3}$)	• Initial weight: 2 g; 8 weeks study • No differences on THC or PO_aw and Ca^{2+} levels • No differences in tissues: HP, HT,
Pacific white shrimp *(Litopenaeus vannamei)*	0; 200^{1}; 500^{2} and $1{,}000^{3}$ µg DON kg^{-1} diet	Trigo-Stockli *et al.* (2000)	• Decrease in performance (↓$W^{3,t1-t4}$, ↓$W^{1,2,t4}$, ↓$Gr^{3,t1-t4}$, ↓$W^{1,2,t4}$)	• Initial weight: 1.7 ± 0.05 g, 16 weeks study (samplings at 4^{t1}, 8^{t2}, 12^{t3} and 16^{t4} weeks) • Naturally contaminated hard red winter wheat
Fumonisin studies				
Pacific white shrimp *(Litopenaeus vannamei)*	0; 200^{1}; 600^{2}; $2{,}000^{3}$ µg FUM kg^{-1} Diet	García-Morales *et al.* (2013)	• Decrease in performance (↓$FW^{2,3}$) • Soluble muscle protein concentration decreased, and changes in myosin thermodynamic properties were observed in shrimp after 30 days of exposure to FB_1. • Marked histological changes in tissue of shrimp fed a diet containing 2.0 µg FB_1 g^{-1} were observed. • Shrimp fed diets containing more than 0.6 µg FB_1 g^{-1} showed greater decrease in shear forces after 12 days of ice storage.	• Initial weight: 5.5–5.7 g, 30 days study

Ochratoxin A studies				
Black tiger shrimp *(Penaeus monodon)*	0; 100[1]; 5,200[2] and 1,000[3] µg OTA kg^{-1} diet	Bundit *et al.* (2006)	• Atrophy, severe necrosis and degeneration of hepatopancreatic tubules, loose contact of hemopoietic tissue and lymphoid organ (HP [T1;3]; HT [T1; 3]; Lyph [T1, 3], HP [T2; 3]; HT [T2; 3])	• Initial weight: 2 g; 8[T1] and 10 [T1] weeks study
Black tiger shrimp *(Penaeus monodon)*	100[1]; 200[2] and 1,000[3] µg OTA kg^{-1} diet	Supamattaya *et al.* (2005)	• Altered physiological parameters (↓ALP[1-3], ↓SGOT[1,2], ↓SGPT[2], ↓PO_aw[3])	• Initial weight: 2 g; 8 weeks study • No differences on THC or Ca^{2+} levels • No differences in tissues: G, AG, HP, HT,

Notes: [a] Studies are ordered alphabetically by common species name.

[b] Red superscript numbers and letters link methods (e.g. specific doses tested) to results of the same study (i.e. presented in the same row).

[c] Purified mycotoxin was applied unless stated otherwise.

Mycotoxins: Afla: aflatoxins (the sum of Afla B_1, Afla B_2, AFG_1 and AFG_2); Afla B_1= aflatoxin B_1; Afla B_2= aflatoxin B_2; AFG_1= aflatoxin G_1; AFG_2= aflatoxin G_2; DON = deoxynivalenol; FUM = fumonisins (the sum of FB_1 and FB_2); FB_1= fumonisin B_1; FB_2= fumonisin B_2; OTA= ochratoxin A; ZEN= zearalenone.

Abbreviations: ALP – Alanine aminotransferase; ALP – Alkaline phosphatase; ALT – Alanine aminotransferase; CAT – Catalase; FBW – Final body weight; FCR – Feed conversion ratio; FR – fat retention; GOT – Glutamic-oxaloacetic transaminase; GPx – Glutathione peroxidase; GR – Growth rate; GSH – Glutathione; GST – Glutathione S transferase; GTP – Glutamic-pyruvic transaminase; HP – hepatopancreas; HT – hematopoietic tissue; Lyph – Lymphocyte; MDA – Malondialdehyde; Nr – Nitrogen retention; PO_aw – Phenoloxidase activity; SGOT – Glutamic-oxaloacetic transaminase; SGPT – Glutamic-pyruvic transaminase; SGR – Specific growth rate; SOD – Superoxide dismutase; W – Weight; WG – Weight gain.

European species sensitive to Afla. When beluga were fed diets containing 25, 50, 75 or 100 μg Afla B_1 per kg for 3 months, weight gain and feed conversion ratio (FCR) were negatively affected (Sepahdari *et al.*, 2010). A broad range of changes in liver tissue were also recorded, including progressive fat deposition, hepatocyte degeneration and necrosis, particularly at concentrations of 75 and 100 μg Afla B_1 per kg diet after 60 days. In another study, juvenile hybrid sturgeon *Acipenser ruthenus x A. baeri* showed increased mortality when fed a diet contaminated with 80 μg Afla B_1 kg^{-1} for 12 days although no external changes or unusual behavior have been observed (Rajeev Raghavan *et al.*, 2011). In the same study, significant histopathological changes including nuclear hypertrophy, hyperchromasia, extensive biliary hyperplasia, focal hepatocyte necrosis and presence of inflammatory cells were observed in the livers of fish fed 40 or 80 μg Afla B_1 kg^{-1}. Furthermore, Afla B_1 accumulation in fish muscle and liver was detected with increasing dietary Afla B_1 levels (Rajeev Raghavan *et al.*, 2011).

While for European temperate species sensitivity levels tend to be relatively low, for tropical species sensitivity levels were observed to vary greatly depending on species but also sometimes within the same species. This variation might be due to a variety of factors as explained before (toxin-related, animal-related and environmental factors). The trophic position of a species may affect its ability to cope with mycotoxins, as it determines whether animals naturally encounter such mycotoxins in their diets (carnivorous versus omnivorous behavior). In one of the most studied and cultivated aquaculture species, Nile tilapia, growth rate and FCR is reported to be significantly affected by Afla B_1 at dietary concentrations ranging from 50 to 2,500 μg kg^{-1} (Chávez-Sánchez *et al.*, 1994; Deng *et al.*, 2010; El-Banna *et al.*, 1992; Hassan *et al.*, 2010; Oliveira *et al.*, 2013; Tuan *et al.*, 2002; Zychowski *et al.*, 2013b). Mortality directly linked to Afla ingestion is also reported to vary greatly. On the one hand, Chávez-Sánchez *et al.* (1994) and Tuan *et al.* (2002) reported no significant effect of dietary exposure to 30,000 μg kg^{-1} Afla B_1 for 25 days in Nile tilapia, while El-Banna *et al.* (1992) observed 16.7% mortality increase, when Nile tilapia were fed only 200 μg kg^{-1} of Afla B_1 for 10 weeks. These differences might be associated to experimental differences and animal-related factors, such as age, sex, or nutritional and immunological conditions before the experiments. For two important Asian aquaculture species, namely rohu (*Labeo rohita*) and gibel carp (*Carassius gibelio*), a higher resistance to Afla was observed than for European carnivorous species. In rohu, Afla B_1 induced oxidative damage to the liver, kidneys and brain when administered at a dose of 100 μg 100 g^{-1} body weight (Madhusudhanan *et al.*, (2004) and immunosuppressive effects when administered at 1,250 μg kg^{-1} body weight (Sahoo and Mukherjee, 2001a, 2001b). Gibel carp showed a fast clearance of Afla B_1 during a recovery period and 12 weeks exposure to up to approximately 1,000 μg Afla B_1 per kg feed showed no effect in this species (Huang *et al.*, 2014). Asian catfish species yellow catfish (*Pelteobagrus fulvidraco*) and Tra catfish (*Pangasius hypophthalmus*) were reported to be sensitive to relatively low levels of Afla (Gonçalves *et al.*, 2018a; Wang *et al.*, 2016).

Effects of Afla were also studied in North American aquaculture species, such as red drum (*Sciaenops ocellatus*) and channel catfish (*Ictalurus punctatus*). Red drum is reported to be highly susceptible to Afla B_1 at levels as low as 0.1 μg kg^{-1} diet. Zychowski *et al.* (2013a) showed that Afla B_1 negatively impacted red drum weight gain, survival, feed efficiency, serum lysozyme concentration, hepatosomatic index (HSI), whole-body lipid levels, liver histopathological scoring and trypsin inhibition. Channel catfish is relatively resistant to Afla B_1 when compared with other species. Jantrarotai and Lovell (1990) reported that gross appearance and behavior of channel catfish were normal after being fed 10,000 μg Afla B_1 per kg^{-1} feed for 10 weeks. However, after 10 weeks, histopathological effects were observed and gastric glands in the stomach were necrotic and contained infiltrating macrophages (Table 3.1).

3.2.2 Deoxynivalenol

Trichothecenes are extremely potent inhibitors of eukaryotic protein synthesis, interfering with initiation, elongation and termination stages (Kumar *et al.*, 2013). Deoxynivalenol (DON) in particular is one of the most frequently found mycotoxins in cereal grains worldwide (Rodrigues and Naehrer, 2012). In aquaculture species, DON ingestion has been associated with a highly significant decrease in growth, feed intake, feed efficiency, and protein and energy utilization (Gonçalves *et al.*, 2018c; Hooft and Bureau, 2017; Hooft *et al.*, 2011; Matejova *et al.*, 2014, 2015; Ryerse *et al.*, 2015). One of the aquaculture species most sensitive to low levels of DON is rainbow trout (*Oncorhynchus mykiss*) (Gonçalves *et al.*, 2018c, 2018e; Hooft *et al.*, 2011; Woodward *et al.*, 1983). Hooft *et al.* (2011) reported that low, graded levels of DON ranging from 300 to 2,600 μg kg^{-1} feed caused a highly significant decrease in growth (–40%), feed intake (–52.7%), feed efficiency (–76.7%) and protein and energy utilization (–74.4% and –72.1%) when compared with the control group that received uncontaminated feed. Recently, Gonçalves *et al.* (2018c) observed that in *O. mykiss* (2.5 g), diets contaminated with 4.7 or 11.4 mg DON kg^{-1} fed for 60 days decreased final body weight, specific growth rate, feed intake, hepatosomatic index and protein efficiency ratio. Furthermore, these fish showed an altered whole-body composition and whole-body nutrient retention. Pepsin activity in stomach samples and lipase activity in intestine samples was increased in fish that received 11.4 mg DON kg^{-1}. Furthermore, fish that received 4.7 or 11.4 mg DON kg^{-1} feed showed increased mRNA expression of insulin-like growth factors 1 and 2 in the liver and of two peptides that regulate feed intake, neuropeptide Y precursor and adenylate cyclase-activating polypeptide, in the brain. Another study with Atlantic salmon (*Salmo salar L.*) found that fish fed 3700 μg DON kg^{-1} showed a 20% reduction in feed intake and a 31% decrease in specific growth rate (Döll *et al.*, 2010). Channel catfish (*Ictalurus punctatus*) fed diets containing up to 10,000 μg DON kg^{-1} either in purified form or contained in naturally contaminated wheat showed no differences in feed consumption, growth, hematocrit values or liver weights compared with animals that received uncontaminated feed (Manning *et al.*, 2014). A feeding trial in carp (*Cyprinus carpio L.*) using three different concentrations of DON (352, 619 and 953 μg kg^{-1}) showed immunosuppressive effects of low dietary DON concentrations (Pietsch *et al.*, 2014) (Table 3.1).

3.2.3 Fumonisins

In aquaculture species, fumonisin B_1 (FB_1) has been generally associated with reduced growth rate, feed consumption and feed efficiency ratio and impaired sphingolipid metabolism (Goel *et al.*, 1994; Li *et al.*, 1994; Lumlertdacha and Lovell, 1995; Tuan *et al.*, 2003). However, information on the effects of FUM on the most important aquaculture species is still scarce. It is known that rainbow trout liver is sensitive to FUM. Fumonisin has been shown to induce changes in sphingolipid metabolism in the liver at levels lower than 100 μg kg^{-1} (Meredith *et al.*, 1998) and to induce liver cancer in 1 month old trout (Riley *et al.*, 2001). Contradictory information is available for Baltic salmon (*Salmo salar*), a species related to rainbow trout. In a feeding trial with Baltic salmon, it was observed that animals that received graded levels of FB_1 (1,000; 5,000; 10,000 or 20,000 μg kg^{-1} feed) for 10 weeks appeared unaffected in terms of growth, feed intake and liver damage (García, 2013). However, in this study it was observed that all fish (including the control group) had very poor appetite and growth, presenting specific growth rates values two to six times lower than the average reported in other studies (Farmer *et al.*, 1983; McCormick *et al.*, 1998). Furthermore, FB_1 concentrations in feed after extrusion were not reported, leading us to assume that real inclusion levels of FB_1 in the feed might be different from the ones theoretically estimated. Adverse effects of FUM contaminated diets have also been reported in carp (*Cyprinus carpio* L.). One-year-old carp showed signs of toxicity with 10,000 μg FB_1 kg^{-1} feed (Petrinec *et al.*, 2004). In these experiments, scattered lesions in the exocrine and endocrine pancreas and inter-renal tissue, probably due to ischemia and/or increased endothelial permeability, were observed. In another study, 1 year-old carp that consumed pellets contaminated with 500; 5,000 or 150,000 μg FB_1 per kg of body weight, showed a loss in body weight and alterations of hematological and biochemical parameters in target organs (Pepeljnjak *et al.*, 2003). For tropical species, Tuan *et al.*, (2003) demonstrated that dietary FB_1 levels of 10, 40, 70 or 150 mg kg^{-1} fed for 8 weeks affected growth performance of Nile tilapia fingerlings. In this experiment, fish fed diets containing FB_1 at levels of 40,000 μg kg^{-1} or higher showed decreased average weight gains. Hematocrit was decreased only in tilapia fed diets containing 150,000 μg FB_1 kg^{-1}. The ratio between free sphinganine (Sa) and free sphingosine (So) in liver increased at a dietary FB_1 level of 150,000 μg FB_1 kg^{-1}. Similar levels of dietary FUM (20, 40 or 60 mg FB_1 + FB_2 kg^{-1}) fed for 30 days were tested by Claudino-Silva *et al.* (2018) in Nile tilapia fingerlings (2.6 g). The authors observed a reduction in weight gain and feed efficiency and reduced mRNA levels of growth hormone receptor and insulin growth factor 1 in the liver, which may be associated with the observed reduction in growth. In channel catfish *(Ictalurus punctatus)* dietary levels of $FB_1 \geq 20,000$ μg kg^{-1} have shown to be toxic (Lumlertdacha and Lovell, 1995) (Table 3.1).

3.2.4 Ochratoxins

Ochratoxins are known for their nephrotoxic and hepatotoxic effects in livestock (Lanza *et al.*, 1980). Furthermore, ochratoxin A (OTA) has already been detected in meat (Guillamont *et al.*, 2005), milk (Skaug, 1999) and dairy products (Dall'Asta *et al.*, 2008) and other animal derived swine products (Pozzo *et al.*, 2010). In aquaculture species, it has been reported that OTA may cause severe degeneration and necrosis of kidney and liver leading to inferior weight gain, poorer FCR, lower survival rates and hematocrit levels (Doster *et al.*, 1972; Lovell, 1992; Manning *et al.*, 2003, 2005b). However, studies on the toxicity of OTA in aquatic animals are very scarce, especially for tropical species. In rainbow trout (*Salmo gairdneri*) fed 2 to 8 mg OTA kg^{-1} of body weight (Doster *et al.*, 1972), severe degeneration and necrosis of kidney and liver, pale kidney, light swollen livers and mortality occurred. The authors report an LD_{50} of 5.53 mg kg^{-1} body weight (13.72 μmols kg^{-1}). In European seabass (*Dicentrarchus labrax* L.), El-Sayed *et al.* (2009) found a 96 h LC_{50} value of 277 μg kg^{-1} body weight with 95% confidence limits of 244 to 311 μg kg^{-1} bw. In addition, common carp (*Cyprinus carpio*) appears to be very sensitive to OTA. Agouz and Anwer (2011) showed that a natural contamination of 15 μg OTA kg^{-1} diet resulted in decreased growth performance and feed utilization parameters. Carcass dry matter, protein and ash contents negatively correlated with OTA. Channel catfish (*Ictalurus punctatus*) that received feed contaminated with 5, 1.0, 2.0, 4.0 or 8.0 mg OTA kg^{-1} showed significantly reduced weight gain, a poorer FCR, lower survival and a lower hematocrit level (Manning *et al.*, 2003). Moreover, moderate to severe histopathological lesions of liver and posterior kidney were observed (Manning *et al.*, 2003). In another study, Manning *et al.* (2005a, 2005b) fed channel catfish diets contaminated with 2–4 mg OTA kg^{-1} for 6 weeks. The authors observed a decreased weight gain and when subsequently subjected to an *Edwardsiella ictaluri* challenge, those fish that had received a dietary OTA level of 4 mg kg^{-1} showed increased mortality (Table 3.1).

3.2.5 Zearalenone

Studies on the effects of zearalenone (ZEN) in farmed animals have mainly focused on dysfunction or structural disorders of the reproductive tract (Minervini and Aquila, 2008; Woźny *et al.*, 2013, 2017; Zinedine *et al.*, 2007). It has been shown that the estrogenic properties of ZEN give rise to a number of reproductive disorders in exposed livestock mammals, including decreased libido, anovulation, infertility or neoplasmic lesions (Minervini and Aquila, 2008; Zinedine *et al.*, 2007). Similarly, in oviparous animals including fish, ZEN mimics the action of natural estrogen, 17β-estradiol, by binding to and activating responsive genes that encode vitellogenin or zona radiata protein – major structural elements of the oocyte (Arukwe *et al.*, 1999; Chen *et al.*, 2010; Woźny *et al.*, 2008). Several studies have confirmed that ZEN modulates estrogen receptor-dependent gene expression affecting the reproduction of fish. This has been shown, for example, in zebrafish (*Danio rerio*) where the exposure to ZEN reduced spawning frequency (Schwartz *et al.*, 2010) or changed the relative fecundity from one generation to the next (Schwartz *et al.*, 2013). In another study, when zebrafish larvae were exposed to ≥ 500 μg ZEN l^{-1}, defects in heart and eye development and upward curvature of the body axis were observed (Bakos *et al.*, 2013). In carp (*Cyprinus carpio* L.), Pietsch *et al.* (2015) investigated the effect of three different dietary ZEN concentrations (332, 621 and 797 μg kg^{-1}) fed for 4 weeks. The authors report no effect on growth, but effects on hematological parameters were observed. An influence on white blood cell counts was noted whereby granulocytes and monocytes were affected in fish fed 621 or 797 μg ZEN kg^{-1} diet. Furthermore, marginal ZEN and α-zearalenol concentrations were detected in muscle samples and the genotoxic potential of ZEN was confirmed by analyzing the formation of micronuclei in erythrocytes. In juvenile rainbow trout (*Oncorhynchus mykiss*), Woźny *et al.* (2012) observed that after 24, 72 and 168 h of intraperitoneal exposure to 10,000 μg ZEN per kg body weight, the mycotoxin interfered with blood coagulation and iron-storage processes. However, in another study with trout, Woźny *et al.*, (2015) observed that a dietary ZEN concentration of 1,810 μg kg^{-1} had no effect on growth, but may have accelerated sexual maturation of female fish (Table 3.1).

3.3 Effect of mycotoxins in shrimp

Impact of mycotoxins in shrimp is less investigated than for fish, which is peculiar taking into account the higher economic interest of these species and the lower number of species being commercially reared in aquaculture.

3.3.1 Aflatoxins

As in fish, Afla are the most studied mycotoxins in shrimp and several reports are available addressing their impact on shrimp homeostasis. Wiseman *et al.*, (1982) reported that 24 and 96 h median lethal doses of Afla B_1 in Pacific blue shrimp (*Penaeus stylirostri*) after injection into the tail muscle were 100.5 and 49.5 mg Afla B_1 kg^{-1}, respectively. Although using intramuscular injection, median lethal doses were quite high. The significance of tested mycotoxin ranges when compared with natural occurring contamination will be addressed in Chapter 4. Regarding the effect of Afla B_1 contaminated feed, black tiger shrimp (*Penaeus monodon Fabricius*) fed Afla B_1 levels ranging from 5 to 20 μg kg^{-1} showed a 46–59% decrease in body weight compared with the control group (Bintvihok *et al.*, 2003). Inconsistently, also in black tiger shrimp, Boonyaratpalin *et al.* (2001) observed that dietary Afla B_1 concentrations ranging from 50 to 100 μg kg^{-1} did not affect growth performance. However, in this study, growth was reduced when Afla B_1 concentrations were elevated to 500–2,500 μg kg^{-1}. The authors observed that survival dropped to 26.32% when 2,500 μg Afla B_1 kg^{-1} diet was given, whereas concentrations of 50 to 1,000 μg kg^{-1} had no effect on survival. In Pacific white shrimp (*Litopenaeus vannamei*), 50 μg Afla B_1 kg^{-1} diet fed for 2 weeks caused abnormal hepatopancreas and antennal gland tissues, 400 μg Afla B_1 kg^{-1} diet significantly affected feed conversion and growth and 900 μg Afla B_1 kg^{-1} diet decreased apparent digestibility coefficients. Tapia-Salazar *et al.* (2017) observed that *L. vannamei* fed a diet contaminated with 75 μg Afla kg^{-1} for 42 days showed decreased weight gain, feed intake and nitrogen retention efficiency compared with shrimp fed control feed. As will be discussed in Chapter 4, 75 μg Afla kg^{-1} is a level of contamination frequently detected in commercial finished feeds. More recently, Zhao *et al.* (2017, 2018) have also studied the effect of Afla B_1 on *L. vannamei*. In a first study, the authors observed that *L. vannamei* fed 15 mg Afla B_1 kg^{-1} feed for 15 days, showed higher mortality and considerable damage in the hepatopancreas. Moreover, several genes related to metabolic functions showed altered expression levels in response to Afla B_1 intake. In a second study, the same concentration of Afla B_1 (15 mg kg^{-1}) fed for 8 days also increased the mortality, caused histopathological changes and increased the activity of antioxidant enzymes. Although the results obtained by Zhao *et al.*, (2017, 2018) are interesting, tested contamination levels are quite high in comparison to levels observed in real farm conditions and therefore, the practical relevance of these observations is unclear. Recently, Yu *et al.*, (2018) fed *L. vannamei* with 500 μg Afla B_1 kg^{-1} for 8 weeks and observed a decrease in weight gain, effects on biomarkers of oxidative stress, increased hepatopancreas enzyme activity in the hemolymph and histomorphological changes (Table 3.2).

3.3.2 Deoxynivalenol

Regarding the impact of DON on shrimp, it was observed that DON levels ranging from 200 to 1,000 μg kg^{-1} diet significantly reduced body weight and/or growth rate in Pacific white shrimp (*Litopenaeus vannamei*) (Trigo-Stockli *et al.*, 2000). Also in *P. monodon*, dietary DON levels of 0.5–2 mg kg^{-1} fed for 8 weeks decreased the specific growth rate, the feeding rate and the activity of liver enzymes in serum (Supamattaya *et al.*, 2005) (Table 3.2).

3.3.3 Fumonisins

Fumonisin B_1 has not been extensively studied as a shrimp feed contaminant. However, the few available studies suggest that *Litopenaeus vannamei* is sensitive to FB_1. García-Morales *et al.* (2013) have shown that white shrimp fed FB_1 at levels from 20 to 200 μg kg^{-1} diet for 30 days showed a reduction in soluble muscle protein concentration and changes in myosin thermodynamic properties. The same authors reported marked histological changes in tissue of shrimp fed a diet containing FB_1 at a concentration of 200 μg kg^{-1} and meat quality changes after 12 days of ice storage, when shrimp received diets containing more than 600 μg FB_1 kg^{-1} (Table 3.2).

3.3.4 Ochratoxins

Ochratoxin A is probably the least studied of the main mycotoxins in the aquaculture community. Nevertheless, Supamattaya *et al.*, (2005) concluded that shrimp feeds occasionally contaminated with OTA up to 1,000 μg kg^{-1} have no negative impact on the shrimp culture industry although altering several physiological parameters. Bundit *et al.* (2006) observed atrophy, severe necrosis and degeneration of hepatopancreatic tubules and loose contact of hemopoietic tissue and lymphoid organs when shrimp were fed 1,000 μg OTA kg^{-1} for 8 weeks (Table 3.2).

3.3.5 Zearalenone

Regarding ZEN, histological changes were observed in black tiger shrimp (*Penaeus monodon* Fabricius) that received feed contaminated with 100; 500 or 1,000 μg ZEN kg^{-1}. Shrimp fed 1,000 μg ZEN kg^{-1} for 10 weeks showed hepatopancreas atrophy, severe necrosis and degeneration of hepatopancreatic tubules and loose contact of hemopoietic tissue (Bundit *et al.*, 2006).

3.4 Combined effects of mycotoxins

Considering that compound feed contains a mixture of several raw materials and, adding to this, that mycotoxigenic fungi are usually capable of producing more than one mycotoxin, it is not a surprise to frequently observe mycotoxin co-occurrence in aquaculture finished feeds (Gonçalves *et al.*, 2017, 2018b, 2018d). Mycotoxin co-occurrence in fish feeds was reported in the past in Egypt (Abdelhamid *et al.*, 1998), the USA (Lumlertdacha and Lovell, 1995), Indonesia (Ali *et al.*, 1998), Nigeria (Omodu *et al.*, 2013), Central Europe (Pietsch *et al.*, 2013), Brazil, and generally for Southeast Asia and Europe (Gonçalves *et al.*, 2017, 2018b, 2018d). Despite the well-documented mycotoxin co-occurrence in aquaculture feed and the awareness that mycotoxin co-exposure may lead to additive, synergistic or antagonistic toxic effects, little is known about the real impact of multi-mycotoxin exposure in aquaculture species. In one of the few studies addressing combined effects of mycotoxins in fish, Carlson *et al.* (2001) observed that dietary FB_1 was not carcinogenic in rainbow trout when fed at concentrations of 3.2, 23 or 104 mg kg^{-1} diet for 34 weeks. However, dietary FB_1 ($\geq$ 23 mg kg^{-1} for 42 weeks) promoted Afla B_1 initiated liver tumors. This result also highlights the importance of long-term exposure for the susceptibility of animals. Carlson *et al.* (2001) suggested that the FB_1 promoting activity in Afla B_1-initiated fish was correlated with disruption of sphingolipid metabolism, suggesting that alterations in sphingolipid signaling pathways are responsible for the tumor promoting activity of FB_1. McKean *et al.* (2006a) studied the combined effects of Afla B_1 and T-2 toxin (T2) in mosquitofish (*Gambusia affinis*), showing an additive effect of the mycotoxins. In case of carp (*Cyprinus carpio*), He *et al.* (2010) studied the individual and combined effects of DON and Afla B_1 in primary hepatocytes and concluded that the toxic effect of the combined mycotoxins was bigger than the effects of single mycotoxins. Taking into consideration studies done for livestock species (Segvic Klaric, 2012), we could infer that mycotoxin interactions may also have negative effects on fish and shrimp. For example, in swine Afla B_1 and OTA showed additive effects according to liver weight and blood chemistry but they were antagonists with regard to the degree of renal cortical interstitial fibrosis and relative kidney weight (Harvey *et al.*, 1995). *In vitro* studies done in the human bronchus epithelial cell line BEAS-2B showed synergistic effects of Afla B_1 and T-2, and Afla B_1 and FB_1. In the human hepatocarcinoma cell line Hep G2, there was an additive effect of Afla B_1 and T-2 and a slight antagonism of Afla B_1 and FB_1 (McKean *et al.*, 2006a, 2006b). The combined effect of OTA and FB_1 was also tested in Caco-2 (human colorectal adenocarcinoma) and Vero (green monkey renal) cell lines. The two mycotoxins caused a synergistic effect, possibly due to their induction of reactive oxygen species (ROS) production (Creppy, 2002). With regard to Afla B_1 and OTA, Afla B_1 was found to be mutagenic with metabolic activation and OTA was not mutagenic (Golli-Bennour *et al.*, 2010). As mycotoxin co-contamination of aquaculture feed is frequently observed, combined effects of mycotoxins in aquaculture species should be more thoroughly investigated in the future.

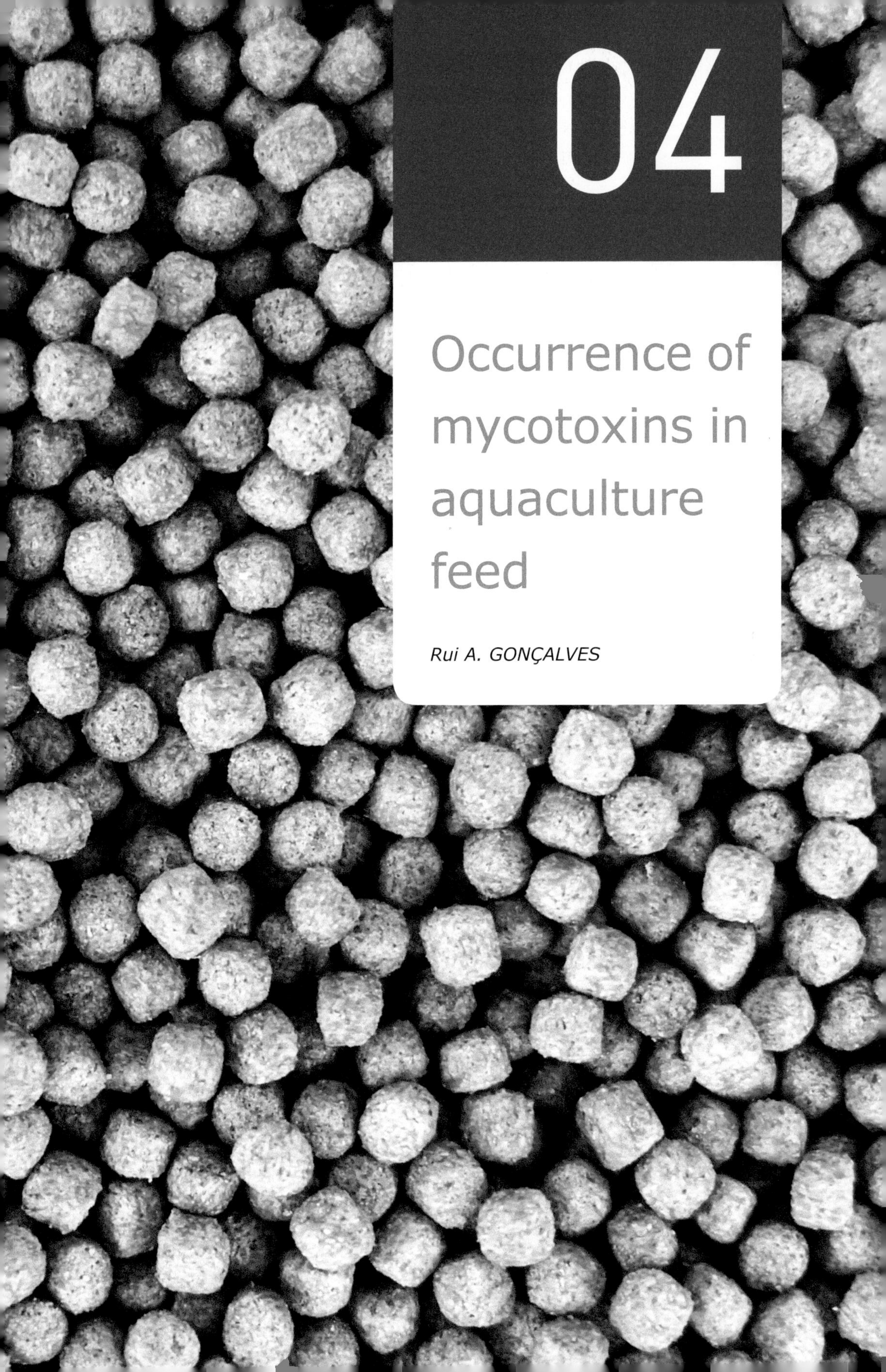

04

Occurrence of mycotoxins in aquaculture feed

Rui A. GONÇALVES

4. Occurrence of mycotoxins in aquaculture feed

Rui A. GONÇALVES

4.1 Aquaculture in an era of finite resources

The growth of the worldwide aquaculture industry has been accompanied by rapid growth of aquafeed production. In 2003, FAO estimated a global aquafeed production of approximately 19.5 million tons and anticipated an increase to over 37.0 million tons by the end of that decade (Rana *et al.*, 2009). In 2017, feed produced for aquaculture was estimated to have reached 39.9 million tons (Reus, 2017). The future growth and sustainability of the industry depends on its ability to identify economically viable and environmentally friendly alternatives to marine derived ingredients, such as fish meal and fish oil. The decreasing supply and high cost of fishmeal have led the industry to concentrate their efforts on finding alternative sources of protein to substitute fishmeal in aquafeed (Davis and Sookying, 2009). Of all the possible alternatives, for example, animal by-products, fishery by-products and single-cell protein, plant-based meals seem to be one of the most promising solutions at the moment (Gatlin *et al.*, 2007).

When considering plant-based meals for aquafeed it is commonly agreed that one of the negative aspects is the presence of anti-nutrients (e.g. cyanogens, saponins, tannins, etc.) that are detrimental to fish and shrimp (Krogdahl *et al.*, 2010). Although there are processes to remove or inactivate many of these compounds, this is not the case for mycotoxins, which are highly stable to processing conditions.

4.2 Selection of plant raw materials for use in aquaculture feed

For most aquaculture species, the selection of plant proteins is based on a combination of local market availability, cost and the nutritional profile (including anti-nutritional factor content and level) of the protein meal in question (Davis and Sookying, 2009; Gatlin *et al.*, 2007; Krogdahl *et al.*, 2010). However, evaluating mycotoxin contamination is not common practice. As clinical signs have not been well investigated (see Chapter 3 for more details), mycotoxins could represent a hidden problem, which can lead to increased disease susceptibility and poor performance. With the overall increase of mycotoxin contamination in plant ingredients and the simultaneous increase in their use, it is important to evaluate the occurrence of mycotoxins in plant proteins, which are commonly used in aquafeed, as well as finished feeds.

4.3 Common mycotoxin-contaminated plant raw materials

4.3.1 Soybean meal

Soybean meal (SBM) is one of the most commonly used plant protein sources in aquaculture feeds, especially in lower trophic level species such as carp, catfish and tilapia, where inclusion levels may exceed 40% (Tables 4.1 and 4.2). Soybean meal is also used for higher trophic level species such as marine fish, salmonids and marine shrimp. However, for these species inclusion levels are normally lower (for more details on inclusion levels for different species consult Tables 4.1 and 4.2). Gonçalves *et al.* (2017) reported that mycotoxins were detected in 88% of Asian SBM samples (n = 48) and in 58% of European SBM samples (n = 19) collected in 2015. The authors reported that Asian samples showed high concentrations of zearalenone (ZEN), deoxynivalenol (DON) and fumonisin (FUM), while aflatoxin (Afla), T-2 toxin (T2) and ochratoxin A (OTA) were found at very low levels. In the European samples, *Fusarium* metabolites were also predominant, with DON showing a maximum concentration of 930 μg kg^{-1}. European samples were also characterized by the presence of T-2 with a maximum value of 105 μg kg^{-1}. Gonçalves *et al.* (2018e) confirmed that SBM sampled in Asia in 2016 is contaminated with the same mycotoxins (i.e. FUM, DON and ZEN), but at different concentrations. Concentrations detected in 2016 were higher when compared with values detected in 2007 (Fegan and Spring, 2007) and in 2015 (Gonçalves *et al.*, 2017). FUM (1,270 μg kg^{-1}) was the main mycotoxin detected in 2016 (average concentrations: fumonisin B_1 (FB_1) 668 μg kg^{-1}, FB_2 = 309 μg kg^{-1}, FB_3 = 294 μg kg^{-1}). The maximum concentration detected for any mycotoxin was 2,571 μg kg^{-1} for FB_1.

4.3.2 Wheat and wheat bran

Wheat (WH) and wheat bran (WB) have a relatively low protein content and lack important amino acids, for example, lysine, threonine and valine, as well as vitamins A and D. However, WH and WB are good sources of phosphorus, vitamin B_1 and other B-complex vitamins. Wheat and wheat bran also have good binding properties, improving the water durability of pellets (Hertrampf and Piedad-Pascual, 2000f). However, the use of WH and WB products in diets for aquatic animals is limited because of the high crude fiber contents. According to Hasan (2007), WH is included at levels up to 20–25% in feed for lower trophic level species. For higher trophic level species, the inclusion level of WH varies greatly with the species and region. For salmonids, an inclusion level of 10 to 14% is common, 15 to 20% for marine shrimps and 5 to 10% for other marine species (Tables 4.1 and 4.2).

Gonçalves *et al.* (2017) reported that 71% of all WH samples (n = 163) collected in Asia in 2015 contained mycotoxins. These samples were characterized by high levels of DON, with an average concentration of 1,275 μg kg^{-1} and a maximum concentration of 6,976 μg kg^{-1}. With 80% of positive samples, DON was also frequently detected in samples collected from Europe in 2015. In European samples, DON showed lower average concentrations (418 μg kg^{-1}) than in Asian samples, but reached a similar maximum concentration (6,219 μg kg^{-1}; Gonçalves *et al.*, 2017). Fumonisin was also detected at relatively high levels in European samples (maximum concentration of 1,628 μg kg^{-1}). Gonçalves *et al.* (2018e) observed that FUM contamination of WH was lower in 2016 than in 2015 for the same commodities and same regions. This highlights the need for constant mycotoxin monitoring and management, as mycotoxin contamination patterns may vary according to several factors such as the origin and quality of the commodities, as well as climatic factors.

Regarding WB, 100% of samples from Asian countries and 80% of European samples were contaminated with mycotoxins in the mycotoxin survey from 2015 (Gonçalves *et al.*, 2017). Asian samples were

***Table 4.1** – Plant ingredients used within feeds for higher trophic level fish/shrimp species*
Source: Gonçalves et al. (2017)

	Plant proteins (%)						
Country	SBM	WH	CGM	R/CM	LKM	FBM	Other plant protein sources
Salmons – Atlantic salmon, coho salmon, chinook salmon							
Norway	8–12	10–14					Others = 20
UK	10	10		5		5	FPM = 3
Trout – rainbow trout, sea trout							
Denmark	12	12					
France	10–15	5–10	5–8				PPM 5–10; SPC 5–10%; FPM = 5–10
Greece	10–35	5–15	5–12	5–10			FPM = 5–10
Norway	8–12	10–14					Others = 20
UK	15	10		5		8	FPM = 3
Marine shrimps – white leg shrimp, giant tiger prawn							
China	10–25				0–20		WH by-products 15–25
India	20–25					1–2	G/PM = 15–20
Marine fishes – barramundi, cobia, cods, groupers, halibuts, seabass, seabreams, tunas, yellowtail							
China	10–25			0–20			WHbp = 15–25
France	15–25	5–10	10–18		5–10		PPM = 5–10, SPC = 5–10
Greece	10–35	5–15	5–12	5–10			PPM = 5–10
Norway							
Spain		1–5	4	7	10		SBC = 5–19; PPM = 5–10
Taiwan	15–25	10–15					
UK	15	10					Others = 10

Notes: Acronyms for plant ingredients stand for: soybean meal (SBM), wheat (WH), corn gluten meal (CGM), rapeseed/canola meal (R/CM), lupin kernel meal (LKM), fava bean meal (FBM) and on other plant protein sources we have: field pea meal (FPM), pea protein meal (PPM), soy protein concentrate (SPC), corn (C), soy lecithin (SL), cassava (CA), rice polishing (RP), groundnut/peanut meal (G/PM), rice bran (RB), broken rice (BR) and by-products (bp).

***Table 4.2** – Plant ingredients used within feeds for lower trophic level fish/shrimp species*
Source: Gonçalves et al. (2017).

	Plant proteins (%)								
	SBM	WH	WB	C	CGM	R/CM	CSM	RB	Other plant protein sources
Carps – grass carp, common carp, crucian carp, catla, rohu									
China	0–25	0–25		0–25		20–40			SbDG = 0–8%
India									G/PM = 30, MC = 10
Catfishes – channel catfish, pangasiid catfishes									
India	10	15–20							G/PM = 30%, MC = 10%
Vietnam	30–60								CA = 20–35
Tilapias									
Taiwan	30–35							10–25	
Vietnam	30–60							20–30	CA = 20–35
Eels									
Denmark	10	15							
Taiwan	8–10	Starch 15–20							
Freshwater prawns									
China			5–10						
India	20–25	20–25							G/PM = 15–20, MC = 15–20
Taiwan	15–20	10–15							

Note: Abbreviations for plant ingredients stand for: soybean meal (SBM), wheat (WH), wheat bran (WB), corn (C), corn gluten meal (CGM), rapeseed/canola meal (R/CM), cottonseed meal (CSM), rice bran (RB) and on other plant protein sources we have: spirit-based distiller's grains (SbDG), groundnut/peanut meal (G/PM), cassava (CA), mustard cake (MC).

contaminated with *Fusarium* mycotoxins, with average values of 620 μg kg^{-1} for FUM, 761 μg kg^{-1} for ZEN and 1,660 μg kg^{-1} for DON. In European samples, FUM was most prevalent and detected at a maximum concentration of 5,334 μg kg^{-1} and DON was detected at an average concentration of 5,124 μg kg^{-1} and a maximum concentration of 15,976 μg kg^{-1}. Average concentrations were considerably higher in WB samples than in WH samples. This can be explained by the tendency of mycotoxins to accumulate in certain cereal milling fractions, depending on their polarity and the type of milling (Cheli *et al.*, 2013). For example, in a study by Mankevičienė *et al.* (2014), detected mycotoxin concentrations in WB were several times higher than in WH.

4.3.3 Corn and corn gluten meal

Corn meal is an excellent energy source but low in protein. It is a major feed ingredient for terrestrial farm animals and its utilization in aquaculture feeds is increasing (Hertrampf and Piedad-Pascual, 2000b). In addition to being low in protein, corn meal is also low in crude fiber. The apparent digestibility of corn meal nutrients varies widely between species, but generally the digestibility can be remarkably improved by gelatinization (Hertrampf and Piedad-Pascual, 2000b). Conversely, corn gluten is the protein portion of the corn kernel and is a by-product of wet milling in the processing of starch. Corn gluten meal (CGM) has a protein content of around 60%. Similar to corn meal, the low levels of amino acids lysine, methionine and tryptophan are limiting factors. Corn is mostly used for lower trophic level species and inclusion levels vary from 10 to 40% depending on the region and species (Table 4.1 and 4.2). By contrast, CGM is mostly used for high trophic level species where inclusion levels are currently below 10% (Tacon *et al.*, 2011) but likely to increase.

According to Gonçalves *et al.* (2017), in a mycotoxin survey conducted in 2015, corn was among the most heavily contaminated commodities sampled in Europe and Asia. In Asia, corn was contaminated with high levels of FUM (average concentration of 2,038 μg kg^{-1}; maximum concentration of 16,258 μg kg^{-1}) and in Europe it was contaminated with high levels of DON (average concentration of 2,469 μg kg^{-1}; maximum concentration of 19,180 μg kg^{-1}). Besides these high levels of DON and FUM, many samples were also contaminated with other *Fusarium* mycotoxins and with Afla. When comparing the concentrations detected in 2015 with previous years (2009 to 2011; Rodrigues and Naehrer, 2012), it can be concluded that FUM have been the dominant mycotoxins in corn from Asia for several years. For European corn samples, we observed a change in the pattern of mycotoxin occurrence over the years. Between 2009 and 2011, FUM was detected at the highest average concentration (2,226 μg kg^{-1}) followed by ZEN and DON and (1,203 and 207 μg kg^{-1}, respectively). In 2015, DON reached the highest average concentration (2,469 μg kg^{-1}), followed by FUM and ZEN (1,462 and 453 μg kg^{-1}, respectively). In 2016, Gonçalves *et al.* (2018e) observed a lower average concentration of DON (418 μg kg^{-1}) in samples from Europe compared with previous years (Gonçalves *et al.*, 2017), but a much higher average concentration of Afla (72 μg kg^{-1} compared with a maximum of 14 μg kg^{-1} in the previous year; Gonçalves *et al.* (2017)). These variations in mycotoxin contamination patterns highlight the need for constant mycotoxin monitoring and management. The mycotoxin survey conducted in 2016 (Gonçalves *et al.*, 2018e), also highlighted that, in addition to the common mycotoxins analyzed in previous studies, other fungal metabolites and masked mycotoxins were detected (nivalenol, 3-acetyl deoxynivalenol, 15-acetyl deoxynivalenol, fusarenon X-glucoside and diacetoxyscirpenolD).

In the mycotoxin survey reported by Gonçalves *et al.* (2017) analyzing aquaculture feed from Asia and Europe sampled in 2015, CGM was the most heavily contaminated commodity. In case of both regions, samples were mainly contaminated with *Fusarium* mycotoxins (ZEN, DON and FUM). This is not surprising, as these were also the main mycotoxins detected in corn and CGM is a corn by-product. All Asian samples and 91% of the European samples contained at least one mycotoxin. Both in 2015 (Gonçalves *et al.*, 2017) and 2016 (Gonçalves *et al.*, 2018e), FB_1 was detected at high levels in CGM (average contamination of 6,107 μg kg^{-1} and maximum of 12,167 μg kg^{-1}). For the sum of FB_1, FB_2 and FB_3 (FB_1 = 6,107 + FB_2 = 2,379 + FB_3 = 599), the average concentration was higher in 2016 than in the previous year (9,085 μg kg^{-1} in 2016; 2,476 μg kg^{-1} in 2015). Both in 2015 and in 2016, all CGM samples were contaminated with mycotoxins.

4.3.4 Rapeseed/canola meal

Rapeseed meal or "canola meal" is a by-product of oil production from rapeseed (*Brassica* spp.). Rapeseed cultivars are low in indigestible carbohydrates and in anti-nutritional factors such as glucosinolates, phenolic compounds (tannins and phenolic acid) and phytic acid (Hertrampf and Piedad-Pascual, 2000e). They have a relatively high-protein content, which is characterized by a well-balanced amino-acid composition and a high biological value. Rapeseed is one of the world's leading edible oil crops. Therefore, the protein rich rapeseed/canola meal (R/CM) is widely available and less expensive than other plant feedstuffs (e.g. soybean meal; Adem *et al.*, 2014). The potential of R/CM as an aquafeed ingredient has been tested in common carp (*Cyprinus carpio*), rainbow trout (*Oncorhynchus mykiss*), turbot (*Psetta maxima*) and tilapia (*Oreochromis niloticus;* Adem *et al.*, 2014). It is generally assumed that the nutritional quality of rapeseed products as a fish meal substitute largely depends on the level of anti-nutritional compounds (Adem *et al.*, 2014). According to Tacon *et al.* (2011), R/CM is already used commercially both for high and low trophic level species, but inclusion levels vary greatly depending on the species and region, e.g. 5% for salmonids in UK and up to 40% for carp in China (Tables 4.1 and 4.2). When analyzing mycotoxin concentrations in a small number of samples from Asia and Europe, Gonçalves *et al.* (2017) observed that R/CM from Europe was mainly contaminated with low levels of DON and FUM (average of positive samples < 45 μg kg^{-1}) and R/CM from Asia was mainly contaminated with ZEN and DON. Values obtained for Asian samples were higher than for European samples and showed a maximum concentration for DON of 2,431 μg kg^{-1}. The only other scientific publication addressing the occurrence of mycotoxins in R/CM (Bojana Kokic, 2009), reported the absence of mycotoxins from one sample of this commodity collected from Serbia. The evidence suggests that in Europe, the risk of mycotoxins in this commodity is low. However, further and more representative analyses should be conducted, especially for other parts of the world.

4.3.5 Cottonseed meal

Cotton (*Gossypum hirsitum*) is grown for its fiber used in the manufacture of textiles. Cottonseed is a by-product of cotton manufacturing. Cottonseed meal (CSM) has relatively high crude protein content. However, the high crude fiber content can be a limiting factor in the use of CSM as feed for aquatic species. Phosphorus, potassium and iron content of cottonseed meal is high and B-vitamin content is more favorable compared with soybean meal (Hertrampf and Piedad-Pascual, 2000a). Because of the low market price in comparison with other plant ingredients and fishmeal, CSM has emerged as a candidate for incorporation in high-protein aqua feeds (Gatlin *et al.*, 2007; Kumar *et al.*, 2013). Nutritionally, it has the advantages of containing high-protein levels and being very palatable for fish (Robinson and Li, 1994). The inclusion of CSM in diets has previously been tested in species such as tilapia (Mbahinzireki *et al.*, 2001; Pavan Kumar *et al.*, 2014; Rinchard *et al.*, 2000), channel catfish (*Ictalurus punctatus*; Robinson and Li, 1994a; Robinson and Tiersch, 1995) and carnivorous species such as chinook salmon (*Oncorhynchus tshawytscha*), coho salmon (*Oncorhynchus kisutch*; LG, 1980) and rainbow trout (*Oncorhynchus mykiss*; Dabrowski *et al.*, 2001). However, according to Tacon *et al.* (2011), CSM is most commonly used for channel catfish and tilapia in USA and China, respectively, at inclusion levels up to 25%. According to Gonçalves *et al.* (2017), only 33% of CSM samples collected from Europe in 2015 were contaminated with mycotoxins. The mycotoxin contaminations were very low and did not represent a threat for aquaculture species. However, in Asia the scenario was completely different and as expected due to the higher temperature and humidity in this region, Afla was the main mycotoxin present (average concentration of 2,038 μg kg^{-1}; maximum concentration of 16,258 μg kg^{-1}). However, *Fusarium* toxins (ZEN and DON) were also found in considerable amounts (Gonçalves *et al.*, 2017).

4.3.6 Rice bran

Rice bran (RB) contains the bran layer and the germ of the rice kernel. It is an inexpensive energy feedstuff that is used in simple fish production systems to grow omnivorous and herbivorous species (Hertrampf and Piedad-Pascual, 2000c). Rice bran is used as a supplementary feed for *Pangasius* by traditional farmers in Vietnam and by Chinese fish farmers in intensive crucian carp culture (Hasan, 2007). According to FAO (2011), RB is most often commercially used for tilapia, at inclusion levels between 10 and 25%, depending on the region. According to Gonçalves *et al.* (2017), RB samples collected from Asia in 2015 were mainly contaminated with ZEN (average value of 147 μg kg^{-1}; maximum value of 545 μg kg^{-1}) and FUM (average value of 118 μg kg^{-1}; maximum value of 713 μg kg^{-1}) and also contained lower levels of DON and Afla. In the following year, DON and ZEN levels detected in samples of this commodity collected from the same region were markedly higher (average concentration of DON: 1,535 μg kg^{-1}; average concentration of ZEN: 515 μg kg^{-1}; Gonçalves *et al.*, 2018e).

4.3.7 Other plant raw materials

The increasing pressure on the use of the above-mentioned crops by both the growing human population and by livestock feed millers, leads to continuously rising costs. This, in turn, is stimulating the use of alternative feedstuffs that are locally available (Lukuyu *et al.*, 2014). Those products and by-products, which are commonly available locally, are non-competitive feedstuffs. They are starting to be developed as components of aquaculture feeds. However, very little information is available on the mycotoxin risk of these materials. In 2016, Gonçalves *et al.* (2018e) studied mycotoxin occurrence in some of these locally available and non-competitive plant feedstuffs used for aquaculture. The authors analyzed samples of alfalfa, cassava, groundnut cake, sesame, sunflower cake and dried distiller's grains with solubles (DDGS) obtained from fermented WH. In addition, aquaculture by-products, such as sun-dried fish and shrimp head meal were investigated (Chapter 4.4). The seven analyzed DDGS samples were contaminated, on average, with seven different mycotoxins. One of the DDGS samples was highly contaminated with ZEN and FB_1 (7,279 μg kg^{-1} and 4,568 μg kg^{-1}, respectively). The remaining DDGS samples were contaminated with DON (average concentration of 1579 μg kg^{-1}) and FUM (average concentration of 823 μg kg^{-1}; sum of FB_1 = 571 μg kg^{-1}; FB_2 = 181 μg kg^{-1}; FB_3 = 71 μg kg^{-1}, Gonçalves *et al.*, 2018e).

The only sunflower sample collected was found to be contaminated with two mycotoxins. Both of them were detected at low concentrations with aflatoxin B_1 (Afla B_1) showing the highest level, i.e. 4 μg kg^{-1}. The only analyzed cassava sample was contaminated with four mycotoxins, all of them were detected at low concentrations. FB_1 showed the highest concentration (33 μg kg^{-1}) in this sample. Alfalfa (n = 1) was contaminated with low levels of five mycotoxins. In this sample, DON showed the highest concentration (151 μg kg^{-1}). Groundnut cake (n = 5) was contaminated with four mycotoxins per sample on average. The highest levels were detected for Afla (average of 376 μg kg^{-1}, maximum of 1,174 μg kg^{-1}). A sesame sample was contaminated with low concentrations of four mycotoxins with ZEN showing the highest level of 35 μg kg^{-1}. The low number of samples collected makes it difficult to draw solid conclusions on the risk that mycotoxin contamination of these plant meals may pose to the aquaculture sector. However, it encourages the frequent monitoring of these plant meals for the presence of mycotoxins.

4.4 Aquaculture by-products

Shrimp head meal is an important by-product of the shrimp industry. It is estimated that 50% of the whole shrimp is commercially processed and the head makes 34 to 45% of whole shrimp (Hertrampf and Piedad-Pascual, 2000d). This by-product is a valuable feedstuff for aquaculture. In some Southeast Asian countries, fishmeal is produced by grinding sun-dried fish. While not being a typical product to analyze for the presence of mycotoxins, it is known that contamination is possible (Fegan and Spring, 2007), especially with mycotoxins such as Afla and OTA, that are produced under poor storage conditions. However, Gonçalves *et al.* (2018e) analyzed two fish samples in the mycotoxin survey conducted in 2016 and detected FB_1 and FB_2, which are mycotoxins produced by *Fusarium* molds on growing plants rather than during storage (Gonçalves *et al.*, 2018e). Interestingly, Fegan and Spring (2007) also reported several marinederived samples of fishmeal and shrimp meal contaminated with mycotoxins produced by *Fusarium* spp. The mycotoxin occurrences in fish and shrimp feeds found by Gonçalves *et al.* (2018e) in 2016, were higher than in previous studies for the same region (Fegan and Spring, 2007; Gonçalves *et al.*, 2017, 2018c). Fegan and Spring (2007) reported several samples of fishmeal contaminated with surprisingly high values of T-2 and ZEN. As explained before, these *Fusarium* mycotoxins are generally produced on crops under field conditions, so their production due to inappropriate storage conditions is unlikely. Fegan and Spring (2007) suggested *Fusarium* strains such as *F. oxysporum* and *F. solani*, known as opportunistic pathogens of fish and shrimp (Hatai *et al.*, 1986; Lightner, 1996; Ostland *et al.*, 1987; Souheil *et al.*, 1999), as a possible source for the contamination. The capacity for *F. oxysporum* or *F. solani* to produce mycotoxins has not been investigated so far. Another hypothesis proposed by Gonçalves *et al.* (2018e) is FUM contamination due to bioaccumulation. Recently, Michelin *et al.* (2017) showed that lambari fish (*Astyanax altiparanae*) fed more than 50 μg of Afla B_1 kg^{-1} feed, presented Afla B_1 in muscle after 120 days at similar levels as in feed. However, bioaccumulation of FUM in fish and shrimp has not been investigated so far.

4.5 Accumulation of mycotoxins in plant feedstuff by-products

The probability of mycotoxin occurrence and co-occurrence as well as the contamination level of mycotoxins in aquafeed will always directly depend on the plant raw materials used and their inclusion levels. As mycotoxins are not destroyed during processing of commodities, an accumulation of mycotoxins along the production chain is expected, particularly in processed ingredients (Cheli *et al.*, 2013). Mycotoxins are known to be concentrated in certain milling fractions. The survey of European and Asian feed samples collected in 2016 and reported by Gonçalves *et al.* (2017) exemplified this accumulation of mycotoxins in certain plant by-products. While WH samples from Europe showed an average DON concentration of 418 μg kg^{-1}, samples of WB from the same region showed a much higher average concentration of 5,124 μg kg^{-1}. Other authors reported that concentrations of mycotoxins in some WH milling fractions may commonly be 150 to 340% higher and, in some cases, up to 800% higher than concentrations in WH grains (Cheli *et al.*, 2013). This is an important topic because of the frequent use of these by-products in aquaculture feeds. Furthermore, it was observed that mycotoxin concentrations in CGM can reach three times the values found in whole corn (Gonçalves *et al.*, 2018e). The accumulation of mycotoxins in by-products also depends on the polarity of the mycotoxin. In the case of corn used for ethanol production, it is known that Afla do not accumulate in ethanol but are concentrated in the DDGS (also used for aquaculture feeds though not analyzed in this study). In wet-mill processing Afla concentrate in the gluten by-products (Khatibi *et al.*, 2014).

4.6 Mycotoxin (co-)contamination of compound feed

Considering that compound feed contains a mixture of several raw materials and that mycotoxigenic fungi are usually able to produce more than one mycotoxin, it is not a surprise to frequently observe mycotoxin co-occurrence. Gonçalves *et al.* (2018d) showed that 50% of analyzed aquafeed samples collected from Europe in 2014 contained more than one mycotoxin. In the same study, 84% of the samples collected from Asia were contaminated with more than one mycotoxin. Surveys of mycotoxin occurrence in fish feed have been conducted previously in Egypt (Abdelhamid *et al.*, 1998), the USA (Lumlertdacha and Lovell, 1995), Indonesia (Ali *et al.*, 1998), Nigeria (Omodu Foluke Olorunfemi *et al.*, 2013), Central Europe and Switzerland (Pietsch *et al.*, 2013) and Brazil (Barbosa *et al.*, 2013). It is known that mycotoxin co-exposure may lead to additive, synergistic or antagonist toxic effects (Alassane-Kpembi *et al.*, 2013). Gonçalves *et al.* (2017), observed that samples of finished feed collected in 2015, showed a lower mycotoxin contamination than samples collected in 2014 (Gonçalves *et al.*, 2018d). While in 2014 average values for Afla, ZEN, DON, FUM and OTA were 52, 60, 161, 173 and 2 $\mu g\ kg^{-1}$, respectively, in 2015, samples from the same region contained 58, 53, 29, 0, 58 and 3 $\mu g\ kg^{-1}$, respectively (Gonçalves *et al.*, 2017). For European feed samples, the decrease in mycotoxin concentrations between 2015 and 2014 was even more marked (Gonçalves *et al.*, 2018d). Nevertheless, while average mycotoxin concentrations decreased from 2014 to 2015, the co-occurrence risk increased. In 2014, 84% of the Asian samples were contaminated with more than one mycotoxin and in 2015 co-occurrence increased to 90%. For European samples, a similar trend was observed, the fraction of samples contaminated with more than one mycotoxin increased from 50% to 75%.

In the survey of mycotoxins in aquafeed samples conducted in 2016, it was found that mycotoxin occurrence patterns in shrimp feeds and fish feeds were slightly different from each other, probably reflecting the type of commodities used for the different species (Gonçalves *et al.*, 2018e). While shrimp feeds were mainly contaminated with DON, a typical mycotoxin found in WH, fish feeds were mainly contaminated with FUM, typical contaminants of maize products. Shrimp feeds were mostly contaminated with only low levels of DON, with the exception of two samples, which were contaminated with 2,287 and 329 μg DON kg^{-1}. All shrimp feeds were also co-contaminated with Afla.

4.7 Summary

Mycotoxins were found in most of the commodities and finished feeds that were analyzed so far, indicating that they represent a risk for the development of the aquaculture sector. These analyzed ingredients were of variable origin and quality and currently, it is difficult to estimate the extent of mycotoxin contamination of aquaculture feeds and feed ingredients. While in some cases, the contamination levels were rather low, in others the contamination levels might represent a risk for aquaculture species. In surveys of aquaculture feed samples from Asia, we observed that SBM, WH, WB, corn, CGM, R/CM and RB were mostly contaminated with *Fusarium* mycotoxins ZEN, DON and FUM, whereas CSM and peanut feedstuffs were mainly contaminated with Afla and *Fusarium* mycotoxins ZEN and DON. European samples were mainly contaminated with *Fusarium* mycotoxins. We furthermore found that mycotoxins frequently co-occurred in these commodities and that plant by-products showed higher mycotoxin concentrations than their source material indicating accumulation of mycotoxins during processing of commodities.

Fusarium mycotoxins were also the main mycotoxins found in shrimp and fish compound feeds in these surveys. Especially for shrimp feeds, some of the concentrations found were the highest reported to date and may represent a serious threat to shrimp production in Southeast Asia. It is difficult to estimate the risk that these concentrations pose to aquaculture species, as most published studies investigating the effect of mycotoxins in aquaculture species tested the effects of considerably higher mycotoxin concentrations. Moreover, there are few studies addressing the possible synergism of mycotoxins in aquaculture species. The high diversity of aquaculture species and factors such as age, nutritional and health status, rearing density and environmental conditions of animals tested, influence the outcome of trials leading to variable sensitivity levels, sometimes even for the same species.

Drawing firm conclusions about the impact of mycotoxins in aquaculture is difficult and much more research is still needed. However, even with the few existing pieces of literature and the knowledge already created around this topic, it is clear that the mycotoxin levels found in finished feeds can be expected to negatively influence the aquaculture industry by affecting growth performance and feed efficiency and by making animals more susceptible to diseases.

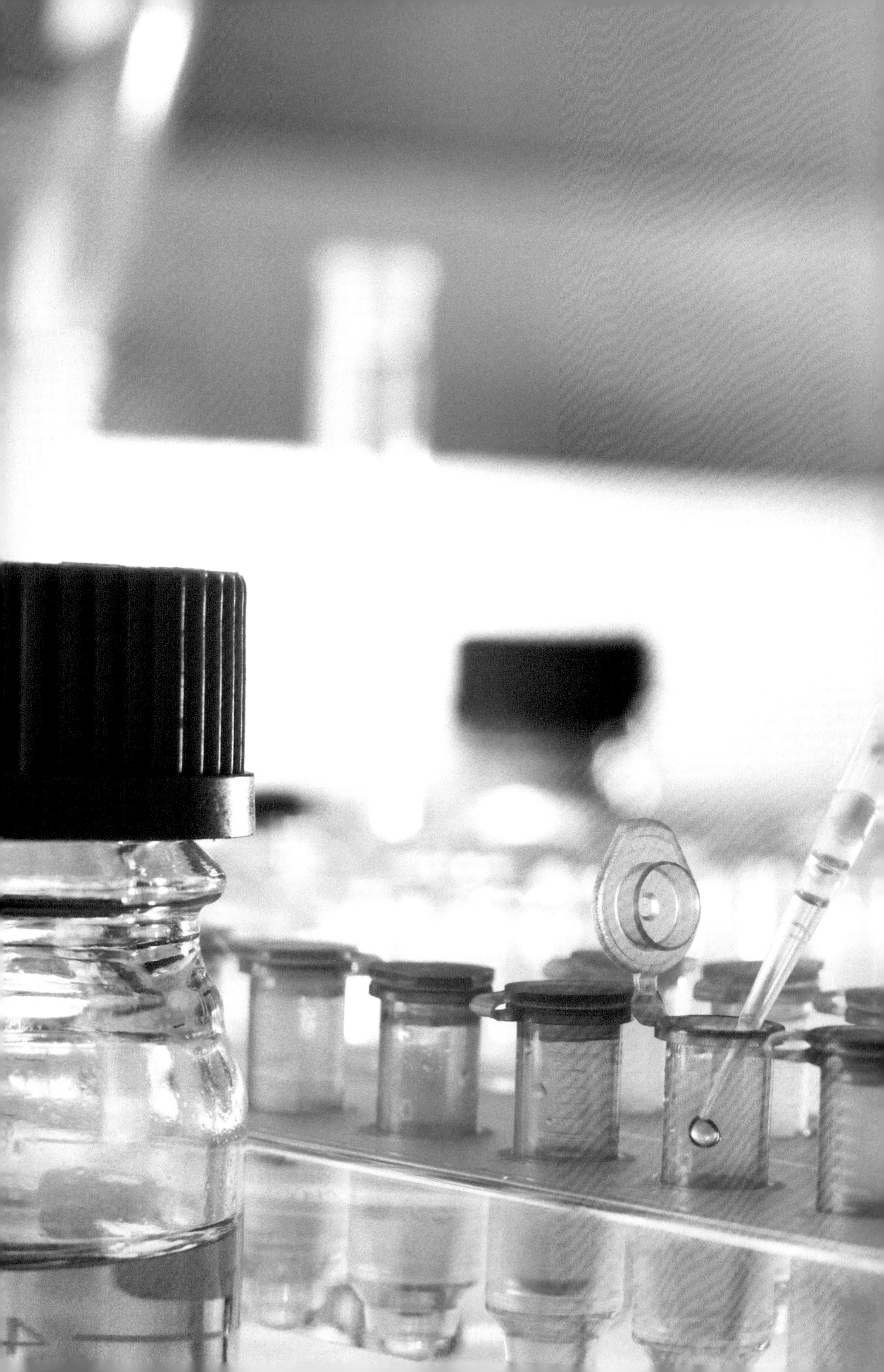

05

Analyzing mycotoxin content in commodities/ feeds

Elisabeth PICHLER and Michele MUCCIO

Edited by Anneliese Mueller

5. Analyzing mycotoxin content in commodities/feeds

Elisabeth PICHLER and Michele MUCCIO

Edited by Anneliese Mueller

The visual diagnosis of mycotoxicoses in animals is complex and often erroneous as different etiologic agents can cause the same symptoms. The best way to identify a problem involving mycotoxins is by analyzing commodities or finished feed for their presence.

Nonetheless, analyzing samples for the occurrence of mycotoxins is not a simple task. A sampling procedure is a multistage process and consists of three distinct phases: sampling, sample preparation and analysis (Cheli *et al.*, 2009). Procedures of handling feed samples comprise the collection of the largest sample size possible and testing samples soon after sampling to avoid changes with respect to quality and contamination.

5.1 Sampling

Sampling of commodities and/or feed is the first critical step concerning chemical analysis of mycotoxins. It is well known as largest source of error (80%) in terms of mycotoxin detection. This is explained by the fact that fungal development and mycotoxin production are "spot processes" significantly affected by crop variety, agronomic practices, weather conditions during growing and harvest, storage and processing conditions and toxigenic potential of the different mold species (Cheli *et al.*, 2009). As these "hot spots" of heavily contaminated material are randomly distributed within a lot, an underestimation of mycotoxin content is possible if a too small sample size without contaminated portions is analyzed. Conversely, an overestimation of mycotoxin load may occur if a small sample featuring one or more contaminated spots is tested (Figure 5.1).

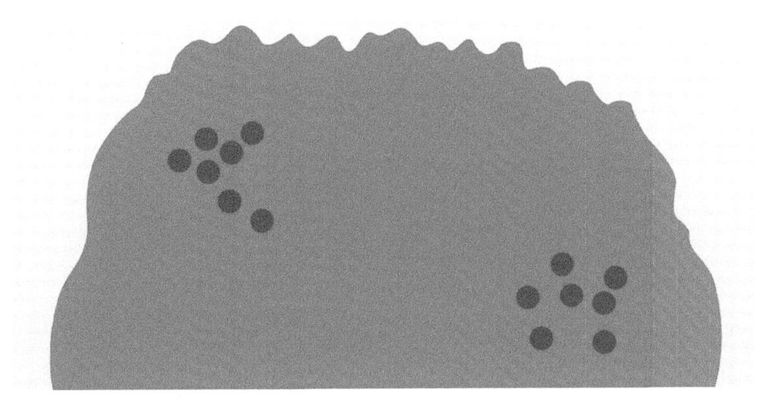

Figure 5.1 *– Inhomogeneous distribution of mycotoxins (dark orange) in grains.*

As an attempt to overcome these problems, several sampling methods/proposals are available for cereals and cereal products, depending on the country.

In the EU, Commission Regulation (EC) No. 401/2006 of February 2006, last amended by (EU) No. 519/2014 in May 2014, lays down the methods of sampling and analysis for the official control of the levels of mycotoxins in foodstuffs. The objective of this regulation is to fix general criteria that the sampling method should comply with. The proposal of the EC will be generally explained below. This can be used as a guideline for managers in the animal production industry with adaptations to individual cases, depending on the available infrastructures.

The method of sampling may be applied to all the different forms in which the commodities are put on the market as commodities may be traded in bulk, containers, or individual packaging such as sacks, bags and retail packages. For the sampling of lots traded in individual packs, such as sacks, bags and retail packages, the following formula may be used as a guide:

$$\text{Sampling frequency} = \frac{\text{weight of lot} \times \text{weight of incremental sample}}{\text{weight of aggregate sample} \times \text{weight of individual packaging}}$$

where weight is given in kilograms and the sampling frequency is every *n*th sack or bag from which an incremental sample must be taken (decimal figures should be rounded to the nearest whole number).

In general, incremental samples of around 100 g should be taken from the lot. The number of samples may vary according to the lot weight (Table 5.1).

If the lot weight is less than 50 tons, the sampling plan is to be adapted with 3 to 100 incremental samples, depending on the lot weight (Table 5.2).

Other crucial parameters to be taken into account in regard to sampling are:

- Sampling raw materials or mixtures of commodities: When deciding whether to test single commodities or mixtures of commodities (e.g. finished feed) one should always bear in mind that the first option enables an easier way to obtain reliable results. It further represents an easier and faster problem solution, as contaminated component can be exchanged or reduced.

***Table 5.1** – Subdivision of lots into sublots depending on product and lot weight*
Source: Commission Regulation (EU) No. 519/(2014) of May (2014).

Commodity	Lot weight (metric ton [MT])	Weight or number of sublots	Number of incremental samples	Aggregate sample weight (kg)
Cereals and cereal products	> 300 and < 1,500	3 sublots	100	10
	≥ 50 and ≤ 300	100 ton	100	10
	< 50	–	3–100 (*)	1–10

Notes: (*) Depending on the lot weight – see Table 5.2. 'Lot' means an identifiable quantity of a food commodity delivered at one time and determined by the official to have common characteristics, such as origin, variety, type of packing, packer, consignor or markings. 'Sublot' means a designated part of a large lot in order to apply the sampling method on that designated part. Each sublot must be physically separate and identifiable. 'Incremental sample' means a quantity of material taken from a single place in the lot or sublot. 'Aggregate sample' means the combined total of all the incremental samples taken from the lot or sublot.

***Table 5.2** – Number of incremental samples to be taken depending on the weight of the lot of cereals and cereal products*
Source: Commission Regulation (EC) No 401/(2006) of February (2006).

Lot weight (MT)	Number of incremental samples	Aggregate sample weight (kg)
≤ 0.05	3	1
> 0.05 and ≤ 0.5	5	1
> 0.5 and ≤ 1	10	1
> 1 and ≤ 3	20	2
> 3 and ≤ 10	40	4
> 10 and ≤ 20	60	6
> 20 and ≤ 50	100	10

- Granulometry or particles and/or seed size: Smaller grain particle sizes usually reflect a better distribution of mycotoxins, so in this case, a smaller subsample may be taken.
- Techniques used to physically collect samples: In case sampling is done in a static manner, i.e. the samples are removed from a static lot (storage bins, rail cars, bags etc.), a probe should be used. It is important that the lot has been well mixed prior to the collection of the sample. As far as possible, these incremental samples should be taken at various places at both surface and depth throughout the lot or sublot, so that every part of the feedlot has an equal chance of selection. Dynamic sampling, though, should be the preferred way of sampling. In this case, samples are removed from a moving stream of product, while this is transferred (e.g. from a conveyor belt). Automatic equipment can be used in this method, to ease the sampling method.
- Tools used for collecting samples: A higher variability of results has been associated if a smaller probe is used (Park *et al.*, 2000). Therefore, the use of a long probe, which will not discriminate the material to be sampled (by taking only big or small particles, for example), is recommended.

5.2 Sample preparation

Sample preparation represents the second crucial step in a proper sampling procedure. It consists of the careful combination of the incremental samples in order to achieve an aggregate sample, which is then usually reduced to obtain a laboratory sample. For this step, two approaches can be used.

- Dry milling: where samples are milled without application of water. This process might lead to clogging of samples with high oil content.
- Slurry mixing: where samples are milled together with an appropriate amount of water at high speed in a slurry mixer, resulting in a homogeneous paste.

In general, the application of slurry mixing technique leads to smaller particle size and a better homogenization of the sample. This will reduce the subsampling variability and enable a better estimation of the true mycotoxin content of a lot (Cheli *et al.*, 2009).

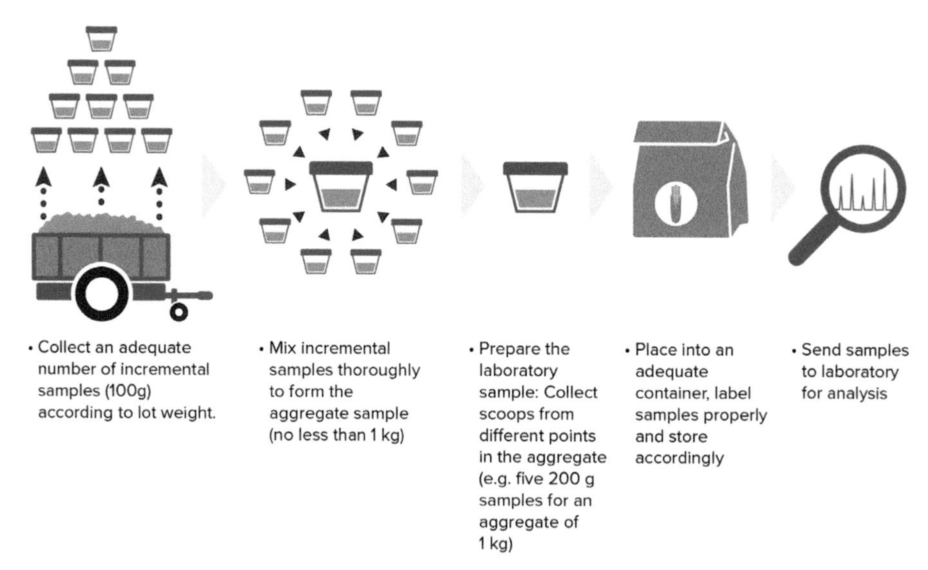

Figure 5.2 *– Schematic of the sampling process.*

If samples are to be analyzed by an external laboratory, these should be placed in an appropriate container. Freezing or airtight packing is necessary for feed samples with high moisture content (liquid feed, silage).

Figure 5.2 shows a simplified procedure prior to analytical testing.

5.3 Analytical methods

The analytical process follows the sampling procedure. Many considerations could be of relevance, but only the more practical ones will be discussed here.

Before analyzing a commodity/feed sample one should ask: "Which result am I expecting from the analysis?" If a yes/no response or a semi-quantitative response is considered satisfactory, then rapid tests (Section 5.3.1) serve the purpose and enable an increase in knowledge of the presence and distribution of mycotoxins in feedstuffs. If a more accurate result is needed, then reference methods (Section 5.3.2) are to be used. An overview of the advantages and disadvantages of rapid methods vs. reference testing methods is given in Table 5.3.

5.3.1 Rapid tests

Rapid tests provide quantitative results in the calibration range and for validated commodities. They require minimal lab equipment and basic personnel training. These are suitable for screening in quality control laboratories in feed and food industries and are commonly used by veterinarians and for on-site testing.

Rapid tests are packaged systems of the principal or key components of an analytical method used to determine the presence of a specific analyte in a given matrix. Rapid test kits include directions for their use and are often self-contained, complete analytical systems. Still, they may require supporting supplies

***Table 5.3** – Overview on the advantages and disadvantages of rapid methods vs. reference testing methods*

	Rapid methods	Reference testing
Advantages	• Fast • Inexpensive • Very reliable for raw materials (corn, wheat) • Quantitative for the validated commodities	• Reliable and quantitative for most commodities • Result refers to the single toxins • Necessary for legal issues
Disadvantages	• Matrix problems can occur in e.g. mixed feedstuffs • Result can be a sum of similar toxins e.g. all Type-B trichothecenes	• More time consuming • Relatively expensive

and equipment. Rapid tests can detect a specific analyte in a given matrix in significantly less time than reference methods (www.aoac.org).

For a rapid detection of mycotoxins, the following technologies are widely accepted by the industry and recognized by the scientific community: enzyme-linked immunosorbent assay (ELISA), lateral flow test and fluorometry.

5.3.1.1 ELISA

ELISA is one of the most popular immunological-based methods used for the analysis of mycotoxins in foods and feeds. The reaction is carried out in 96-well microtiter plates. The compound of interest (e.g. aflatoxin B_1 [Afla B_1]) present in the sample delivered to the laboratory, reacts with specific antibodies attached to the surface of the reaction plate wells. These antibodies are designed to bind Afla B_1. In the reaction, the Afla B_1 contained in the sample solution competes for the antibodies with a known amount of the same mycotoxin (Afla B_1) that is purposely added to the reaction well. This known Afla B_1 is labeled by the manufacturer with a molecule that produces a detectable signal, usually a color change, when properly excited by a specific liquid solution. The sample solution and the labeled mycotoxin are added to the reaction wells and allowed to compete for the antibodies for a certain period of time (usually a few minutes). Afterwards, the wells are washed to eliminate nonbound mycotoxins, and a special liquid solution that excites the molecular label is added to produce the color. As a result, the more Afla B_1 is present in the sample, the lighter the color will be, as only a small amount of labeled Afla B_1 will bind to the antibody. Vice versa, if the sample does not contain Afla B_1, the color will be darker, as more labeled Afla B_1 will bind to the antibody. The procedure is illustrated in Figure 5.3. Results can be influenced by matrix interferences and by a possible cross-reaction of the antibodies. ELISA is only suitable for validated matrices.

5.3.1.2 Lateral flow test

Lateral flow tests consist of relatively simple technology based on a series of capillary beds, such as pieces of porous paper. The first element (filter pad) acts as a sponge and sucks in the sample solution. The fluid migrates to the second element (the gold pad) where the manufacturer has installed the bio-active particles (conjugate): a special dry matrix designed to guarantee a chemical reaction between the

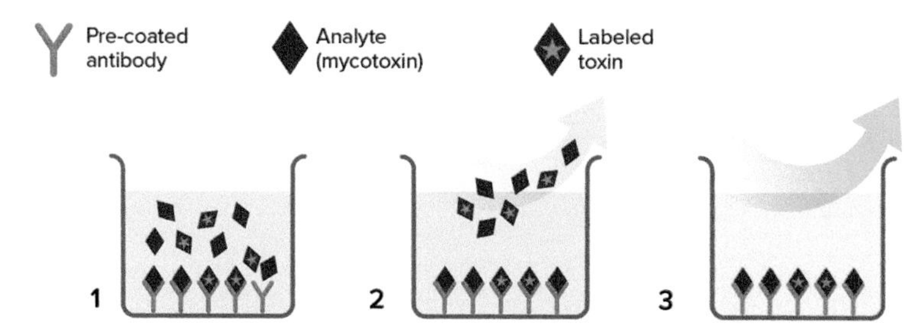

***Figure 5.3** – A typical ELISA reaction. The mycotoxin contained in the sample solution competes with a known amount of labeled mycotoxin added to the reaction (steps 1 and 2). After washing away the non-bound elements, a special liquid solution that excites the molecular label is added to produce the color (step 3).*

target molecule (e.g. mycotoxin) and its chemical partner (antibody) immobilized on the gold pad surface. As the sample solution diffuses up the stripe, it comes into contact and reacts with the matrix containing the antibody on the gold pad. The target molecule (e.g. mycotoxin) binds to the antibody while migrating further through the membrane, toward the adsorbent pad. The membrane has one or more areas (referred to as strips) where a third 'capture' molecule has been installed by the manufacturer. By the time the sample-conjugate mix reaches these strips, the third 'capture' molecule binds the complex. As more and more fluid passes over the strips, particles accumulate and the strips change color. Typically, there are at least two strips. The control strip captures any particle, thereby showing that the reaction conditions and technology are working. The second strip contains a specific capture molecule designed to capture only the sample-conjugate complex (Figure 5.4). Matrix interferences might influence the results.

5.3.1.3 Fluorometry

The basis of fluorometry is the quantification of compounds by measuring their fluorescence using a fluorometer. In some cases, the compounds may be innately fluorescent, and in others the compounds

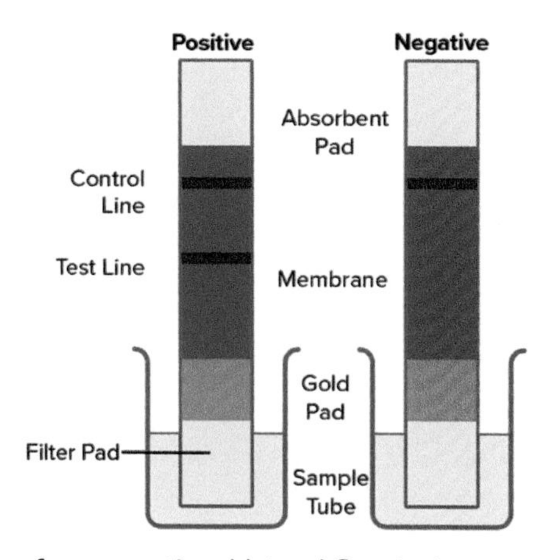

***Figure 5.4** – Illustration of a conventional lateral flow test.*

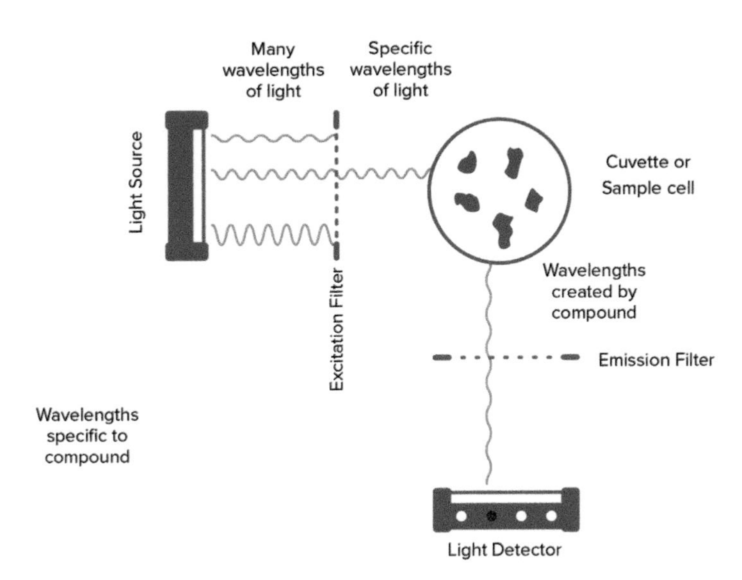

***Figure 5.5** – A sample is scattered with UV light and the emitted wavelength – specific to the sample – is measured.*

are rendered fluorescent by chemical derivatization. Tests for compounds using fluorometry include the extraction of the compound from the specific matrix, followed by a clean-up process using either immunoaffinity columns or solid phase clean-up columns, and then a derivatization (if necessary) and measurement of the fluorescence (Figure 5.5).

5.3.2 Reference testing

A reference method is an internationally or nationally recognized standard analytical method for a particular analyte being tested ("Association of Analytical Communities", AOAC International at www.aoac.org).

Reference testing is more widely used in legal issues, as in the case of governmental control and enforcement of legislation, and in the food industry. It requires expensive lab equipment and well-trained personnel. The advantages of using this quantitative method is the achievement of low detection limits, the possibility of testing very complex commodities (for instance finished feed and silage) and the analysis of several substances at once. ISO 17025 is the main standard used by testing and calibration laboratories. ISO 17025 accreditation ensures that a laboratory is audited at least once a year by strict auditors who check not only whether processes are in place but also people are competent and that constant improvement is attained.

The most common analytical reference methods for the detection of mycotoxins are described below.

5.3.2.1 Gas chromatography

Gas chromatography uses sophisticated equipment in which compounds are separated by a gas flowing through a heated glass column coated with a stationary nonvolatile liquid. Samples injected into the system are separated into the specific components on the column and the separated analytes coming off the column are detected by a chemical or physical detection system (Figure 5.6).

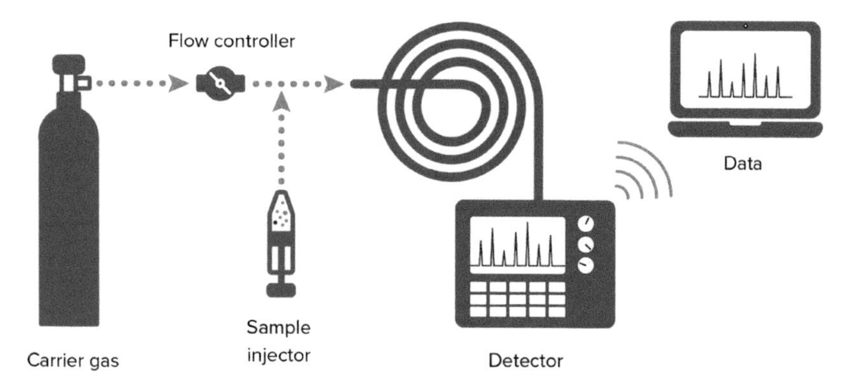

***Figure 5.6** – A sample is carried by a gas into a heated glass column coated with a non-volatile liquid. Different substances will cross the column at different rates. The different substances will generate peaks that are read by the detector and shown on the computer.*

5.3.2.2 High performance liquid chromatography

High performance liquid chromatography (HPLC) is a chromatographic technique, where a small portion of a sample to be analyzed is injected into a stream of solvent being pumped through a column of an adsorptive matrix. As the sample moves through the column the compounds are separated based on the interactions between the solvent system (mobile phase) and the matrix (stationary phase). The sample components elute off the column as separate entities, and the flow of solvent with the respective compounds passes through a detector to measure the response of the specific compound. The concentration of each analyte of interest is calculated by comparing the measured signal with the signal given by standard solutions that were injected into the instrument as part of the analysis (Figure 5.7).

HPLC can be coupled with a variety of detectors, for example, spectrophotometric (UV-VIS) detectors, refractometers, fluorescence detectors, electrochemical detectors, radioactivity detectors and mass spectrometers.

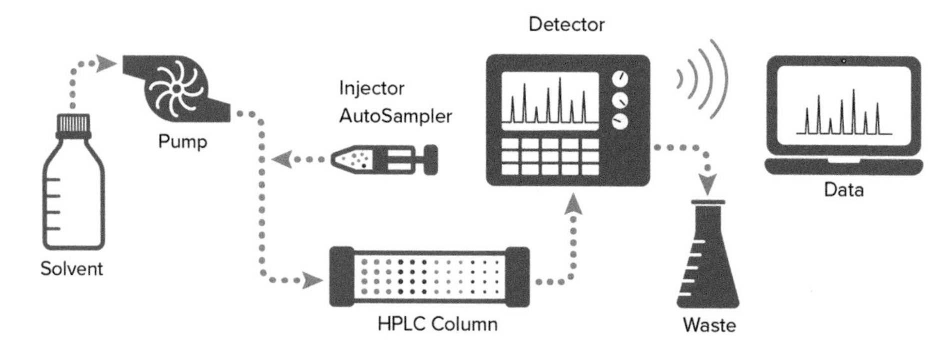

***Figure 5.7** – Typical HPLC reaction where a carrier liquid carries the sample of interest into the column. Different components separate at different rates.*

5.3.2.3 Liquid chromatography-mass spectrometry

The state-of-the-art technology of LC-MS or LC coupled to tandem mass spectrometry (LC-MS/MS) allows efficient spectrometric assays in a routine laboratory setting. This technique combines the physical separation proprieties of the HPLC with the mass analysis capabilities of the mass spectrometer. The two analytical methods work synergistically. Chromatography separates mixtures with multiple components (e.g. mycotoxins), before the mass spectrometer then provides the structural features of individual components with high sensitivity and specificity.

In LC-MS/MS, the mass to charge ratio of the ions belonging to individual mycotoxins are measured before then being fragmented. Each fragment is re-measured in the second mass spectrometry step for extra specificity (Figure 5.8). Owing to the extreme sensitivity, this method is the reference method of choice in many laboratories and it currently represents state-of-the-art of analytical chemistry.

This technique can be used for a wide range of potential analytes, with no limitations by molecular mass, a very straightforward sample preparation and not requiring chemical derivatization. The main advantages include low detection limits, the ability to generate structural information, the requirement of minimal sample treatment and the possibility to cover a wide range of analytes. Sometimes co-eluting matrix components influence the ionization efficiency of the analyte positively or negatively, impairing the repeatability and accuracy of the analytical method. Therefore, a sample clean-up prior to liquid chromatography will be necessary. In order to overcome matrix effects, stable isotope labeled internal standards are used.

5.3.3 Spectrum 380®

Spectrum 380® is a next generation LC-MS/MS technique that enables the simultaneous detection of more than 380 fungal metabolites in feed raw materials and in finished feed. Apart from the so called "major" mycotoxins – compounds like Afla or deoxynivalenol whose toxicological properties have been well characterized – hundreds of other fungal metabolites can be routinely monitored with this method. In a survey of 1,059 feed samples collected from all over the world, major mycotoxins accounted for

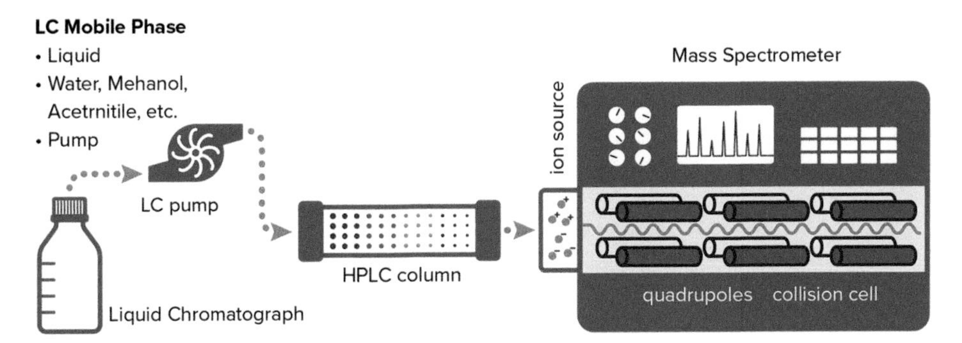

***Figure 5.8** – The LC-MS/MS combines the separation capabilities of the HPLC with the ability to detect the mass and other structural features of a specific sample. In the tandem mass spectrometry (MS/MS) the ions of the individual components are measured at first, then fragmented and each fragment is re-measured in the second mass spectrometry step, for extra molecular specificity and sensitivity.*

merely 5 of the 37 most commonly detected fungal metabolites. Many of the other frequently detected compounds are known as so called "emerging" mycotoxins, that is, fungal metabolites whose toxicological potential in animals is unclear. Also, two frequently detected compounds namely deoxynivalenol-3-glucoside and zearalenone-sulfate represent modified forms of major mycotoxins ("masked" mycotoxins) that may be converted to the respective toxic compound inside the digestive tract. Routine multiple mycotoxin analysis of feed samples enables the identification of compounds that animals regularly encounter and can consequently indicate compounds whose toxicity and mode of action should receive more scientific attention. Interestingly, co-occurrence of fungal metabolites in feed seems to be a common phenomenon. In 96% of the analyzed samples, 10 or more compounds have been detected indicating that animals are typically exposed to a high number of fungal metabolites simultaneously. Toxic interactions of these co-occurring compounds have to be considered in order to understand the effect of a given diet on animal health and they are receiving increasing scientific attention (Alassane-Kpembi *et al.*, 2016; Grenier and Oswald, 2011). Additionally, with the Spectrum 380® method, phytoestrogens, pesticides and veterinary drugs can be detected in the samples.

5.3.4 Spectrum Top® 50

Spectrum Top® 50 uses the Multi-Mycotoxin Analysis 50+, an innovative LC-MS/MS method capable of determining over 50 different mycotoxins and metabolites simultaneously with high specificity and sensitivity. It includes the analysis of certain undetected masked mycotoxins. Other than Spectrum 380®, Spectrum Top® 50 provides information on the most frequently occurring and emerging mycotoxins on the field. The effects of all analyzed toxins are known. The output of this analysis is more focused and the time interval until results are available is shorter. The Spectrum Top® 50 method was developed by scientists of Romer Labs, a leading global supplier of diagnostic solutions for food and feed safety, and sister company of BIOMIN.

5.4 Acknowledgements

Parts of this text were originally published in *Mycotoxins in Swine Production* (BIOMIN Edition) and written by Elisabeth Pichler and Inês Rodrigues, supplemented with parts written by Michele Muccio ("Mycotoxin Compendium", BIOMIN GmbH) and edited by Anneliese Mueller.

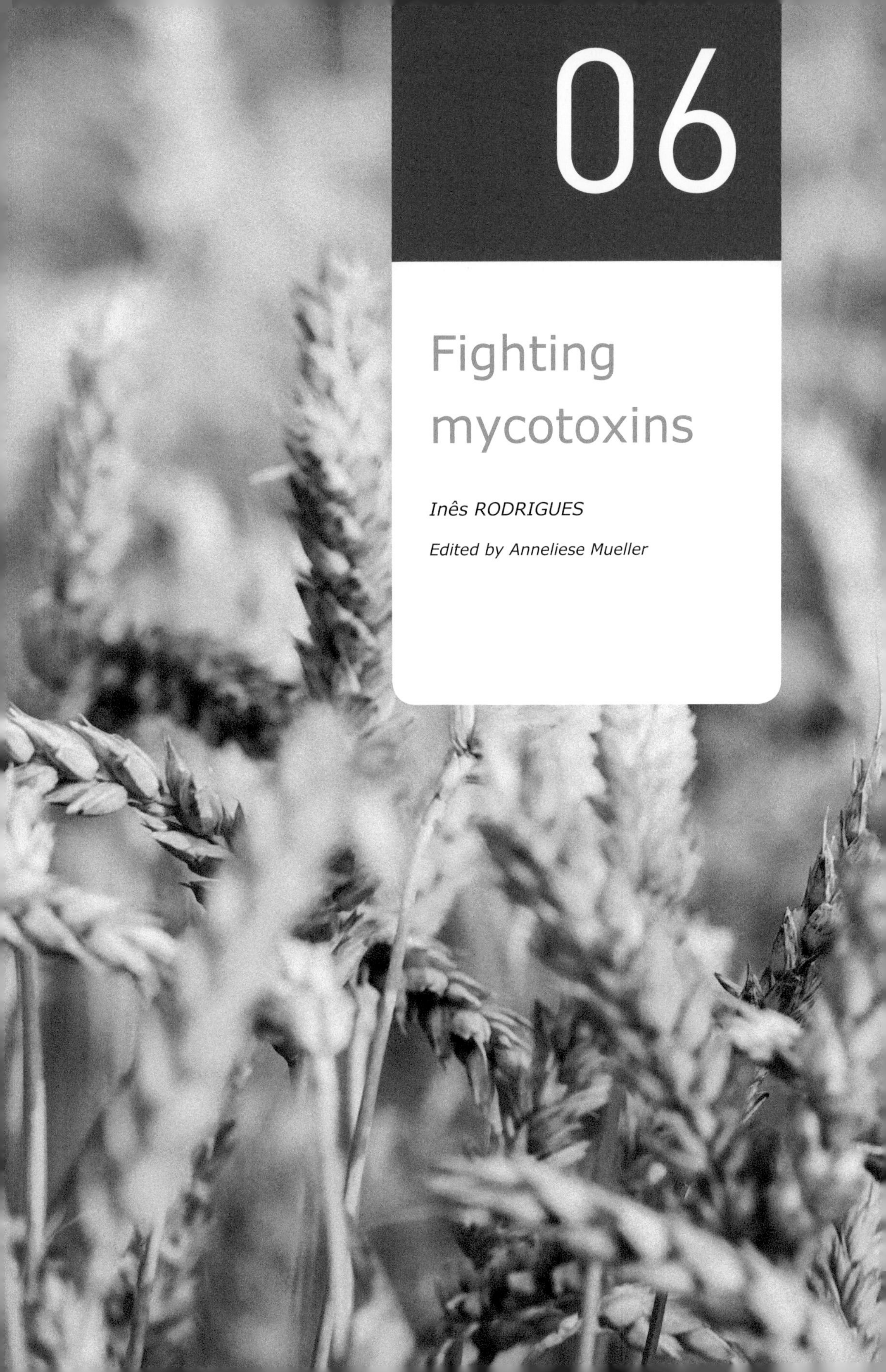

06

Fighting mycotoxins

Inês RODRIGUES

Edited by Anneliese Mueller

6. Fighting mycotoxins

Inês RODRIGUES

Edited by Anneliese Mueller

6.1 Prevention methods

As explained in Chapter 1, proper conditions for growth of fungi can occur at all times during crop growth, harvest and storage. Nevertheless, fungal species are roughly divided into field fungi and storage fungi. In general, field fungi require a higher moisture to grow and to produce mycotoxins (> 0.9 water activity). They infect seeds and plants in the field and mainly belong to the genus *Fusarium*. Storage fungi require a lower water activity and are more prominent after harvest and during storage, for example, *Aspergillus* and *Penicillium* spp.

Therefore, prevention of mycotoxin production must already start in the field prior to seeding and should continue until animals ingest the produced feed. Pre-harvest measures and post-harvest storage optimization remain the most effective strategies to prevent mycotoxin contamination of feed (Peng *et al.*, 2018). Nevertheless, they can only reduce but not completely prevent contamination.

6.1.1 Avoiding contamination in the field and during harvest

Many *Fusarium* species infect corn and cereal grains such as *F. graminearum*, *F. culmorum* and *F. verticillioides*. *F. graminearum* is the major causal agent of head blight in cereals and of red ear rot in corn. There are two typical routes of *F. graminearum* infection in corn (Reid *et al.*, 1999): (1) the spores are already on the field when corn silks emerge and the silk channels get infected; (2) birds, insects or extreme weather conditions damage the kernels before their hardening, which provides an opportunity for fungal invasion (Figure 6.1). Rainfall during the silking period is an important factor for infection with *Fusarium* and rainfall at flowering time (anthesis) has a high impact on the occurrence of cereal head blight. Similarly, rainfall has a high impact on infection with *Alternaria* spp. Heavy rain before harvest promotes colonization of cereal grain ears (Los *et al.*, 2018).

Many factors can influence the growth of *Fusarium* fungi and the occurrence of fusariotoxins in the field. In the following some methods aiming to prevent fungal infection will be discussed (Jouany, 2007; Ogunade *et al.*, 2018; Luo *et al.*, 2018; Peng *et al.*, 2018).

- **Plant variety:** A lot of work has been invested in plant breeding over the past few years as it is considered to be the best solution for *Fusarium* control in susceptible crops. However, quantitative trait loci for mycotoxin resistance are often linked with genes encoding morphological plant characteristics; therefore, an improvement to the first trait will usually lead to adverse effects on the agronomic properties of the plant. Additionally, public and scientific doubts on the utilization of GMOs should be considered (Ogunade *et al.*, 2017).
- **Land management/crop rotation:** Cropping systems in which wheat is grown in the same field each year or is rotated with corn in the same field in consecutive seasons, seem to increase the occurrence

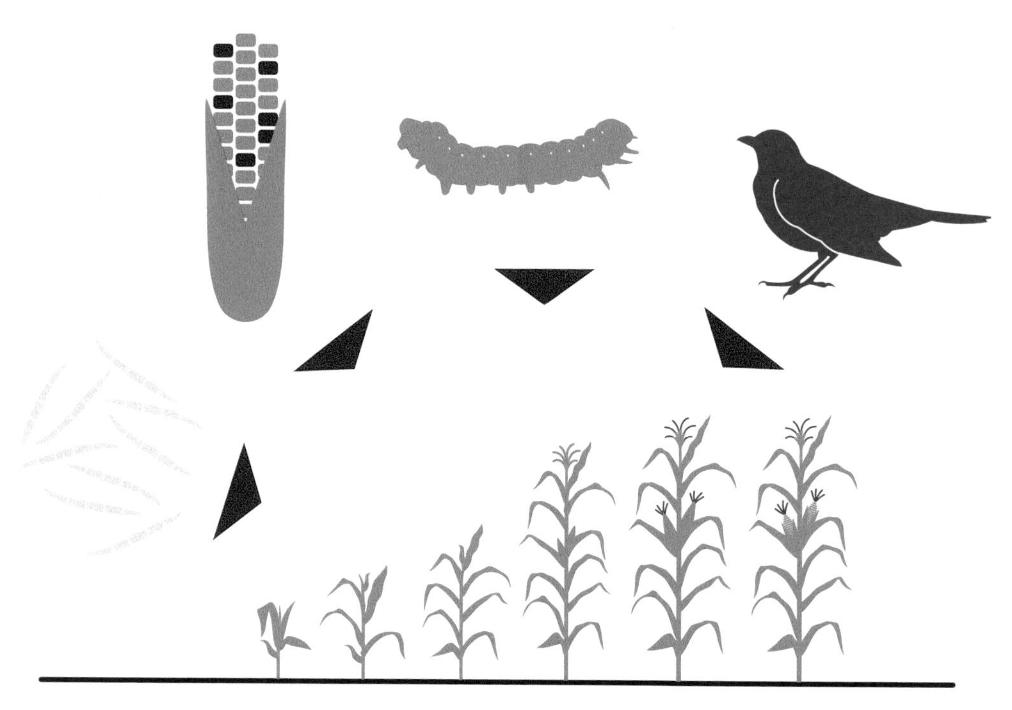

***Figure 6.1** – The typical entry routes of Fusarium graminearum infection in corn.*

of fungal disease. Contrary, a rotation of wheat and legumes has been reported to control mycotoxin contamination (Peng *et al.*, 2018). In a study by Schaafsma *et al.*, 2001, the plant grown in the same field two years ahead could explain some of the variation in concentration of deoxynivalenol (DON) in wheat: if previously other plants than wheat were sowed, the level of DON decreased. Corresponding to this, Codex Alimentarius (2003) suggested crop rotation with a crop not commonly infected with *Fusarium* species (Luo *et al.*, 2018). A careful chosen type of rotation can thus have a positive impact on preventing fungal infection.

- **Planting date:** Planting date/sowing time can influence the risk of *Fusarium* infection. There is a high risk of infection when the plant`s flowering stage is reached at the same time as spores are released from fungi in the soil or on other plants (Jouany, 2007). Less contamination has been observed for early planted maize and for winter varieties of wheat and barley which develop earlier than spring varieties. However, adverse weather conditions can lead to higher contamination of these early varieties.
- **Tillage procedures:** Many *Fusarium* fungi species are soil-borne and survive in crop residues as saprophytes breaking down plant residues. They form spores that are very resistant to adverse temperature and weather conditions, thus facilitating survival in the crop residues. Therefore, it is advisable to eliminate crop residues from the field by means of deep tillage. Deep tillage is more effective than minimum tillage. In the case of no tillage, seeds are directly introduced into the previous crop residues. The idea of no tillage is based on efforts to prevent soil erosion (Dill-Macky and Jones, 2000).

- **Plant stress:** Stress factors, such as high temperatures, drought, poor fertilization and high competition for nutrients are some of the aspects known to increase mycotoxin production in the field. Choosing an appropriate variety of seeds for a certain location, irrigation in critical periods and balanced fertilization are some of the measures that can help to avoid mycotoxin contamination during plant growth.
- **Crop damage:** Mechanical damage or damage by insects or birds to grains provide a good opportunity for fungal invasion and development. Insects are believed to be vectors of fungal spores (Ogunade *et al.*, 2017). Furthermore, the plant's ability to resist mycotoxins might be impaired (Peng *et al.*, 2018). Thus, prevention of crop damage is of major importance.
- **Pesticides:** Pesticides and fungicides can help to prevent damage of the crops by insects and fungal infection, but their usage implicates problems such as health concerns of consumers and development of resistance to pesticides (Ogunade *et al.*, 2017). Consequently, the aim is rather a reduction in the usage of pesticides and fungicides.
- **Bio-agents:** Another approach is the use of "bio-agents". The application of atoxigenic *Aspergillus flavus* or *A. parasiticus* on crops can reduce aflatoxin (Afla) contamination and is used in the US. The atoxigenic bio-agents are supposed to replace the toxigenic ones (Ogunade *et al.*, 2017).
- **Harvesting:** An adequate harvesting date is extremely important to avoid fungal growth and subsequent mycotoxin contamination, as unstable weather conditions such as late rain will dramatically increase its incidence. Early harvest usually leads to less contaminated grains. The time of day also influences the water content of the crop and consequently the development of *Fusarium*. Water content of grains is influenced by atmospheric humidity, which is higher during early morning due to morning dew, than later during the day when the sun is shining (Jouany, 2007). As *Fusarium* spores are known to be prevalent in soil, a contamination with soil should be avoided during harvesting by adjusting the cutting height accordingly (Ogunade *et al.*, 2017). Drying after harvest is another important factor (Awad *et al.*, 2010; Jouany, 2007). Attention must be paid to appropriate harvest equipment. Equipment should be cleaned from previous harvests to avoid cross-contamination and damage of kernels should be avoided as this can promote fungal infection (Awad *et al.*, 2010; Jouany, 2007).

These measures, however, cannot completely prevent mycotoxin contamination. Moreover, after harvest, other fungi can develop and produce mycotoxins depending on the storage conditions.

6.1.2 Avoiding contamination during storage

During storage, precautions should be taken in order to avoid fungal growth and mycotoxin production (Table 6.1).

***Table 6.1** – Factors that influence mycotoxin occurrence during storage and preventive measures*

Factor	Mode of interaction	Preventive measures
Moisture and temperature	Moisture and temperature are the most important physico-chemical factors affecting preservation of commodities and feeds during storage. Both should stay as low as practicable. Nonetheless, fungi can still develop at low temperatures (5 to 10°C) if the moisture level is sufficiently high (Smith and Henderson, 1991). Temperature fluctuations (e.g. rise of 2–3°C) may be a sign of microbial growth and/or insect infestation. These fluctuations can be detected with distributed temperature recorders (Jouany, 2007). Harvested grains usually have a high moisture and drying is one of the most important measures to prevent mycotoxin contamination (Los *et al.*, 2018). A water activity smaller than 0.7 should prevent fungal growth (Los *et al.*, 2018; Peng *et al.*, 2018) and ensure good storage. Generally, a total water content over 150 g/kg in cereals is reported to be necessary for fungal activity. 140 g/kg seem to be enough for *Aspergillus* sp., but *Fusarium* sp. requires at least 170 g/kg. (Jouany, 2007) Typically, moisture content after drying varies between 10–14% (Los *et al.*, 2018). Insufficient drying before storage makes other measures less effective. A critical factor in drying is to guarantee its uniformity despite of the many layers of grains (Los *et al.*, 2018). Rotation of grain can reduce the occurrence of wet spots (Jouany, 2007).	Make sure storage facilities are dry and present low temperatures as well as minimum temperature fluctuations.
Aeration	Mold growth usually occurs heterogeneously in grains, therefore the development of "hot spots" (areas in which the concentration of mycotoxins is higher) is common. This happens because warm air within the grains and feed encounters cooler air, which leads to condensation of water. These high-moisture spots are stimulating fungi germination (Smith and Henderson, 1991). The aim of aeration is to cool the grain. Maintaining it in constant movement will also increase the efficiency of storage. However, it has to be noted that grain dust is hygroscopic – it often has a higher moisture content – and carries a higher proportion of fungal spores than whole clean grain. Fungal development and production of mycotoxins can be further restricted by low oxygen and high carbon dioxide concentration. (Luo *et al.*, 2018).	When possible, aerate the stored goods by circulation of air to maintain proper and uniform temperature levels throughout the storage area. Maintain a low oxygen and high carbon dioxide concentration.

Factor	Mode of interaction	Preventive measures
Sanitation	Fungal development is likely to occur at several points of the storage to feeding pathway, in the storage bins, the feed mill, mixed feed bins, pipelines of the feeding system and ultimately in animal feeders.	Cleaning of equipment on a regular basis is highly recommended.
Pests	Pests are generally injurious or unwanted organisms. Besides fungi, insects can negatively influence storage. Through their metabolic activity temperature and water content can increase (Peng *et al.*, 2018). Furthermore, they cause physical damage of grains and commodities. Arthropods are also carriers of mold spores and their fecal material can be a food source for molds (FAO, 2003). Bacteria do not play an important role in spoilage (Los *et al.*, 2018). Growth of field fungi can also occur if drying was delayed (Los *et al.*, 2018), but problems are rather due to storage fungi (as outlined in the first chapter). Use of fungistatic agents is common and can be very efficient in reducing mold growth and mycotoxin production (Leeson and Summers, 1991). Nevertheless, it should be emphasized that once the mold has already damaged grain and/or produced mycotoxins, the effectiveness of this practice is limited. It is known that sub-lethal applications of fungicides stimulate mycotoxin formation. This is likely to occur because fungi are stressed, but not killed (Benbrook, 2005). Consequently, if fungicides are used, the product instructions should be carefully followed to avoid this adverse effect.	Use good hygiene procedures to minimize levels of insects and fungi in storage facilities. Chemicals used should not interfere with the intended end use of the goods.
Physical damage	Although most mold pathogens can directly penetrate plant tissues, it is important to avoid mechanical and insect damage. Broken kernels caused by handling and/or insect damage facilitate infection of grains (Santin, 2005).	Effective post-harvest insect management and correct equipment calibration is highly recommended.
Filling and feed-out of Silos	Silos should be filled rapidly, packed tightly, and sealed completely to provide anaerobic conditions (Jouany, 2007). Once silos are opened, conditions won't be anaerobic anymore and growth of fungi might occur. Thus, feed-out of silos should be rapid and a straight silo face should be maintained. It's best to feed silage right after removing it (Jouany, 2007). Long term storage and mixing of grains increases contamination risk (Luo *et al.*, 2018). Biological additives, e.g. lactic acid bacteria, can improve silage fermentation, increase aerobic stability and decrease the pH thereby helping to prevent fungal growth (Jouany, 2007).	Fill silos quickly and tight and seal completely. Empty silos rapidly and maintain a straight silo face. Feed silage immediately.

6.2 Elimination of mycotoxins

Preventive methods during crop growth, harvest and storage can only reduce but not eliminate the risk of mycotoxin contamination, thus detoxification procedures after harvest are still necessary. These procedures should deactivate, destroy or remove the toxin and the fungal spores while retaining the nutrient value and acceptability of the feed by the animal. The deposition of toxic substances, metabolites or toxic by-products in the feed and significant alterations in the feed's properties has to be avoided.

Furthermore, detoxification methods should be readily available, easy to use, inexpensive and environmentally friendly. Detoxification procedures are divided into three categories: (1) physical methods; (2) chemical; and (3) biological methods (adsorption and biotransformation).

6.2.1 Physical processes

Many physical processes are aimed at decreasing the mycotoxin contamination of commodities, for example cleaning, mechanical sorting and separation, dehulling, washing, density segregation and thermal inactivation.

The efficacy of these physical treatments depends on the level of contamination and the distribution of mycotoxins throughout the grain. Results are often unreliable and high feed losses may occur. Furthermore, these methods require a high investment of money, thus their practical application is very limited. Nonetheless, they will be briefly referred to below.

6.2.1.1 Cleaning

Broken kernels are more easily infected by fungi than intact kernels and therefore more likely to contain mycotoxins. In addition, grain dust may transport fungal spores, which increases the chance of mycotoxin production. The main objective of the cleaning process is to remove contaminated grain dust, husks, hair and small particles by aspiration or scouring.

6.2.1.2 Mechanical sorting and separation

In this process, the clean product is separated from mycotoxin-contaminated grains. High feed losses are possible due to incomplete and unreliable separation. Therefore, mechanical sorting and separation is not cost effective.

6.2.1.3 Dehulling

The outer layer of grains is mechanically removed. This process decreases the mycotoxin content significantly, but leads also to a high product loss. Depending on contamination levels, hulls can be re-used in nutrition providing a fiber-rich supplement (Peng *et al.*, 2018).

6.2.1.4 Washing

Washing procedures using water or sodium carbonate reduce mycotoxins in grains (Wilson *et al.*, 2004). Nevertheless, this process may be used only in feeds or commodities undergoing wet milling or ethanol fermentation, otherwise costs of drying would be prohibitive.

6.2.1.5 Density segregation

Flotation can segregate contaminated grain and thus lead to a reduction in mycotoxin contamination. However, it should be noted that appearance and weight of a particular kernel does not necessarily indicate mycotoxin contamination.

6.2.1.6 Thermal inactivation

As explained in Section 1.4, mycotoxins are very heat stable and heat treatments such as boiling in water, roasting, pelleting or even autoclaving (for some types) cannot adequately destroy them.

6.2.1.7 Irradiation

In this process, ionizing radiation is applied, which leads to damage to the DNA and metabolic pathways in cells resulting in cell lysis (Los *et al.*, 2018). Some experiments have been carried out applying irradiation to reduce the mycotoxin load of commodities (Aziz and Moussa, 2004; Kottapalli *et al.*, 2003; Ritieni *et al.*, 1999). The results showed a reduction in fungal spore contamination but not in terms of mycotoxins already present in the material. A reduction of the concentration of Afla, ochratoxin A (OTA), T-2 toxin (T2) and DON using irradiation has nevertheless been reported by some studies (Luo *et al.*, 2018). This reduction in mycotoxin concentration is mainly due to heating and hydrolysis (Peng *et al.*, 2018).

6.2.1.8 Further physical decontamination processes

Several other physical treatments have been investigated for their future potential, for instance microwave treatment, where electromagnetic waves are used to inactivate microbes. Here, mainly the thermal effect leads to cell death. A modern nonthermal technology is pulsed UV light treatment, which uses short-duration, high power pulses of UV light and kills cells and spores. However, its application on the surface structure of cereals might not be possible. As last example, nonthermal (cold) plasma should be mentioned. Plasma is one of the four states of matter. It is partially or fully ionized gas. During electrical discharge radicals are generated (e.g. reactive oxygen species) and decontamination is due to these radicals. Application of nonthermal plasma has already been tested on cereal. However, the effect of these different novel approaches on mycotoxins which are already present in the crops, is unknown (Los *et al.*, 2018).

6.2.2 Chemical processes

Various chemicals (acids, bases, aldehydes, bisulfites, oxidizing agents and different gases) have been tested for their ability to detoxify mycotoxins, but only a limited number of chemical methods is effective and may be used in practice. Attention must be paid to a possible formation of toxic residues, negative effects on the nutrient content, flavor, color, texture or functional properties of the feed.

To achieve adequate decontamination efficiency, several parameters such as reaction time, temperature and moisture have to be monitored. Chemical methods need additional cleaning steps and are therefore very expensive and time consuming. Treatment of contaminated feed with ammonia was the most common method used in the past. It leads to a reduction of mycotoxins as well as an inhibition of mycotoxigenic fungal growth (Luo *et al.*, 2018). Although early studies showed that this technique is safe and effective, the use of ammoniation to decrease the Afla level in specific commodities is only permitted in certain countries. Ozone treatment is another chemical method approved for the use in food processing. Its effect is based on its high oxidizing ability (Luo *et al.*, 2018; Peng *et al.*, 2018).

In 2006, the European Commission already banned the application of chemical treatment in food and feed in the Comission Regulation (EC) No 1881/2006 (Peng *et al.*, 2018).

6.2.3 Mineral and biological methods – adsorption

The use of adsorbent materials is a very common approach to prevent mycotoxicoses, in particular aflatoxicosis. These compounds are added to the feed to bind the toxin in the gastrointestinal tract during the digestive process, resulting in a reduction of toxin bioavailability.

The efficacy of mycotoxin binding is dependent on the molecular structure and physical properties of the sorbent as well as on the physical and chemical properties of the mycotoxins (Daković *et al.*, 2005). Substances scientifically investigated as potential mycotoxin-binding agents include bentonites, zeolites, organophilic clays, activated charcoal and yeast cell walls.

6.2.3.1 Bentonites

Bentonites, montmorillonite-rich negatively charged clays, are phyllosilicates characterized by alternating layers of tetrahedral silicon $[SiO_4]^{-4}$ and octahedral aluminum connected in a 2 : 1 arrangement. They occur worldwide.

Bentonites with montmorillonite as the dominant mineral are called smectites. Smectites have a large surface and a high cation exchange capacity, thus polar and cationic organic substances can be adsorbed (Vila-Donat *et al.*, 2018). Various studies investigated the adsorption mechanism of Afla by these phyllosilicate minerals (Kannewischer *et al.*, 2006; Phillips *et al.*, 2002). Adsorption on the surface, at the edges as well as in the interlayer space has been proposed. Deng *et al.* (2010) showed evidence that multi-layer adsorption on external surface of the smectite was unlikely. They proposed that Afla molecules occupy the interlayer space together with exchange cations and water molecules. In the same experiment, the intensity of the bonding was shown to be very high, suggesting high stability of interlayer-adsorbed Afla. High intensity bonding is based on chemisorption (see Section 6.2.4). Bentonites showed the most promising results as far as aflatoxin B_1 (Afla B_1) adsorption is concerned (Vekiru *et al.*, 2007). They are also considered as safe and efficient additives by EFSA (Vila-Donat *et al.*, 2018).

6.2.3.2 Zeolites

Zeolites are porous tectoaluminosilicates that possess an infinite three-dimensional cage-like structure. These negatively charged minerals are characterized by their ability to lose and adsorb water without damaging their structures. They have a large internal surface and high cation exchange capacity (Vila-Donat *et al.*, 2018). This structure allows the retention of molecules smaller than the pores. Due to their selective binding according to shape, charge and size of molecules, they are referred to as molecular sieves (Vila-Donat *et al.*, 2018). However, as stated by Daković *et al.* (2005), the Afla B_1 molecule is too large to enter the zeolite channels, so the adsorption is limited to the external surface of the zeolite particle.

6.2.3.3 Organoclays

The permanent negative charge of the above-mentioned minerals' structure (bentonites and zeolites) makes them suitable for modification using long-chain organic cations (surfactants). This modification results in increased hydrophobicity of the mineral surface and a potentially higher affinity for hydrophobic mycotoxins (Daković *et al.*, 2007). Such organoclays are synthesized by modifying bentonites (or zeolites) with specific organic cations, precisely quaternary amines, in a process named organophilization. Quaternary amines are surfactants with a hydrophilic (water loving) and a lipophilic (oil loving) group. A wide variety of surfactant molecules can be used to modify clays. These compounds vary in number and in size of the alkyl chain groups and identity of the polar head (Lemke *et al.*, 1998). In simple terms, these molecules are placed between the clay layers (in the case of bentonites) to increase the lipophilicity and the distance between layers, thus increasing the adsorption capacity of the clay. Hereby, the cations of the bentonite or zeolite are exchanged for the added organic cations with alkyl chain groups. In the case of zeolites, however, the surfactants are too large to enter the zeolite channels or to access internal cation exchange positions and their adsorption is limited to the external surface of the zeolite.

Because of their high adsorbing capacity, organoclays are for example used in the treatment of wastewater. Adsorption of zearalenone (ZEN), OTA and fumonisin (FUM) has been reported by some studies (Vila-Donat *et al.*, 2018). Lemke *et al.* (1998) studied the interactions of ZEN with organoclays and agreed with previous studies that the greater the hydrophobicity of the clay, the higher the affinity to bind ZEN and the smaller the desorption rate. In other words, adsorption of ZEN steadily increases as the number of exchanged surfactants as well as the length of the surfactants' alkyl chains (> 12 carbon alkyl chain) increases. However, alkyl chains longer than 16 carbon atoms showed an increased desorption rate of surfactants. The behavior and specificity of adsorption of organoclays in complex media (with other nutrients) still has to be fully addressed in scientific studies.

In studies performed at the Department for Agrobiotechnology (IFA-Tulln) of the University of Natural Resources and Life Sciences (Vienna, Austria), the adsorption of FUM by organoclays was reduced by approximately 80% (pH 3 or pH 6), when complex medium was used instead of buffer solution (Table 6.2). The adsorption of OTA was reduced by approximately 20% and 60% at pH 3 and pH 6, respectively (Tables 6.2 and 6.3). This dramatic reduction in FUM adsorption capacity in complex

***Table 6.2** – Organoclays: reduction of fumonisins adsorption due to unspecific binding of feed ingredients (e.g. nutrients, vitamins)*

Fumonisin adsorption (%)			
in buffer solution		in complex medium[1]	
pH 3.0	pH 6.0	pH 3.0	pH 6.0
94.7	43.8	14.8	6.0

Note: [1]Containing essential nutrients, trace elements and vitamins (simulation of the feed).

***Table 6.3** – Organoclays: reduction of ochratoxin A adsorption caused by unspecific binding of feed ingredients (e.g. nutrients, vitamins)*

Ochratoxin adsorption (%)			
In buffer solution		In complex medium[1]	
pH 3.0	pH 6.0	pH 3.0	pH 6.0
83.8	57.1	68.0	22.9

Note: [1]Containing essential nutrients, trace elements and vitamins (simulation of the feed).

media may be due to the non-specificity of the binding between mycotoxins and organoclays. This lack of specificity represents a problem for the use of organoclays as feed additives, as they may adsorb other feed components such as vitamins, nutrients, trace elements and pigments.

A study on the adsorption of Afla in to organoclays found that Afla B_1 is more highly adsorbed by unmodified zeolite than by the organic version of this mineral. This observation may be explained by the almost planar and not very flexible structure of Afla B_1. Furthermore, it has a higher dipole moment than ZEN and ochratoxins (Daković *et al.*, 2005).

To investigate whether organoclays are safe and do not negatively affect health, the toxicity of organoclays was tested in hydra (*Hydra vulgaris*), a well-established and sensitive *in vivo* indicator of toxicity (Marroquín-Cardona *et al.*, 2009). Toxic effects on hydra could be observed, which is in accordance with previous studies in mice (Afriyie-Gyawu *et al.*, 2005; Lemke *et al.*, 2001a). Furthermore, according to IPCS INCHEM (International Programme on Chemical Safety – Chemical Safety Information from Intergovernmental Organizations) quaternary ammonium compounds can cause toxic effects by all routes of exposure including inhalation, ingestion, dermal application and irrigation of body cavities.

6.2.3.4 Activated charcoal

Activated charcoal is formed by pyrolysis of organic materials. It is a very porous non-soluble powder with a high surface to mass ratio (500–3500 m^2/g). It has been shown to have a high affinity for different mycotoxins and to adsorb them effectively in *in vitro* experiments (Vila-Donat *et al.*, 2018). Still, effective adsorbance of the various mycotoxins was not confirmed in *in vivo* experiments. This might be due to the non-specific binding of a variety of different substances including essential nutrients, particularly if these substances are present in the feed at higher concentrations than the mycotoxins (Huwig *et al.*, 2001). Accordingly, Afla B_1 adsorption by charcoal was greatly affected in real gastric juice when compared with the results obtained in buffer solutions and adsorption of vitamin H by activated charcoal reached 99% (Vekiru *et al.*, 2007).

6.2.3.5 Yeast and yeast cell wall derived products

Yeast and yeast cell wall derived products have also been used as adsorbents for mycotoxins. Two major components of the yeast cell wall are glucomannans and mannanooligosaccharides. They are, especially β-D-glucan, directly involved in the physical binding of mycotoxins (Peng *et al.*, 2018). Adsorption of several mycotoxins on the cell wall surface has been reported (ZEN, OTA, FUM) (Vila-Donat *et al.*, 2018). The adsorption of Afla B_1 by glucomannan-based products was found to be lowest at pH 2 and 6.5, as represented by the sorption isotherms (see Section 5.2.4) (Marroquín-Cardona *et al.*, 2009). Reports show that the main forces involved in the molecular mechanism of the binding of Afla B_1 by glucomannans are Van der Waals attractions and hydrogen bonds (Yiannikouris *et al.*, 2006). However, it is known that these bonding forces are reversible, largely dependent on the molecule's orientation (Maroquín-Cardona *et al.*, 2009) and weaker than the bonding forces of chemisorption, which are involved in the binding of Afla B_1 to smectite (Grant and Phillips, 1998). Furthermore, toxicity of yeast products was observed in the hydra toxicity bioassay, which may be due to the growth of intact yeast and microorganisms in these products (Kannewischer *et al.*, 2006; Maroquin-Cardona *et al.*, 2009). These polysaccharides have been reported to have positive side effects limiting the harm of mycotoxins, such as effects on immune activities and gastrointestinal pathogens (Peng *et al.*, 2018).

6.2.3.6 Other adsorbents

Further substances were tested as potential adsorbents of mycotoxins. One example is the polymer cholestyramine, an insoluble quaternary ammonium exchange resin, which has been described to be effective in some studies, but is expensive. Different strains of lactic acid bacteria have been found to bind Afla B_1 and ZEN. Binding is based on cell wall peptidoglycans, polysaccharides and teichoic acids. Furthermore, fibers from plants, mainly composed of (hemi)cellulose and lignin, showed adsorption capacity, although less than other tested substances. Other interesting examples are apple pomace that contains high amounts of fibers and pectin, and grain pomace that has a high content of phenolic compounds (Vila-Donat *et al.*, 2018). A totally different approach is to use magnetic materials and nanoparticles as for example carbon nano-composites to remove Afla B_1 or surface-active maghemite nanoparticles to remove citrinin. These approaches show potential but are still in development (Luo *et al.*, 2018). Also, these approaches might raise health concerns of consumers.

6.2.4 Testing mycotoxin binders

It is still unclear which mineral properties influence mycotoxin adsorption. Aflatoxin adsorption values correlated neither with the amount of smectite in bentonites, nor with their cation exchange capacity (Vekiru *et al.*, 2007). Therefore, thorough tests of potential mycotoxin binders are crucial. Lemke and co-workers (2001) developed a multi-tiered approach for the *in vitro* prescreening of clay-based enterosorbents. Later on, Vekiru *et al.*, (2007) investigated various adsorbents for their ability to bind Afla B_1 with a similar protocol. The protocol includes the following tests.

- Adsorption test for screening at different pH-values – to understand the ability of the materials to bind mycotoxins at different pH conditions.
- Chemisorption test – to evaluate the strength and thus efficacy of the binding, given by the chemisorption index (Cα). A Cα=1 indicates total binding, with no desorption of Afla B_1 from the binder material. The adsorption rate after extensive washing steps is calculated. Chemisorption is based on electron sharing leading to a surface complex of the adsorbent and the adsorbate (Phillips *et al.*, 2002; Vekiru *et al.*, 2007). This is a much stronger bond than physical adsorption based on the rather weak Van der Waal's forces, which can be easily reversed.
- Comparison of adsorption in buffer, artificial gastric juice and real gastric juice – to determine the influence of incubation medium on Afla B_1 adsorption.
- Isothermal analysis – to evaluate the affinity and maximum capacity of materials. The number of bound molecules is influenced by many parameters including temperature, pressure, and, importantly, the surface of the adsorbent, pore volume, ratio of meso- and micropores, surface energy distribution, etc. The adsorption isotherm is the relationship between the quantity of molecules adsorbed and the changing pressure at constant temperature.
- Vitamin binding – comparison of the amount of vitamin and toxin bound to the adsorbent when they are both present in the test solution.

These tests determine whether the material enables chemisorption (which indicates strong binding and no or low desorption of the already adsorbed mycotoxins) and whether it has a high adsorption capacity and a high mycotoxin affinity (with no or low adsorption of essential nutrients). The suitability of binding materials for the use as feed additives has to be further investigated. Essential properties of feed additives are the absence of toxicity (including the absence of heavy metals and dioxins), a low effective inclusion rate in feed, rapid and uniform dispersion in the feed during mixing and heat stability during pelleting, extrusion and storage.

6.2.5 Biological methods – biotransformation

Mycotoxin binders are useful for removing Afla from animal feed, but they are not effective in preventing toxic effects of *Fusarium* mycotoxins, such as trichothecenes or ZEN (Avantaggiato *et al.*, 2005; Huwig *et al.*, 2001) (Figure 6.2). For the elimination of mycotoxins that cannot be removed by mycotoxin binders, biotransformation methods have been developed. Biotransformation is the conversion of mycotoxins into less or non-toxic molecules by microorganisms or their purified enzymes. Microorganisms or enzymes are added to the feed and enable the degradation of mycotoxins in the gastrointestinal tract (Rodrigues *et al.*, 2009).

The concept of biotransformation and studies on this topic go back to the 1960s when the first bacterial strain able to degrade Afla was discovered (Ciegler *et al.*, 1966). Many other scientific studies are available that report mycotoxin converting microorganisms (Alberts *et al.*, 2006; Luo *et al.*, 2018; Ogunade *et al.*, 2017; Varga *et al.*, 2000; Wegst and Lingens, 1983). However, for their use as feed additives, mycotoxin detoxifying microorganisms and enzymes have to meet several demands.

- The formed metabolites have to be non-toxic.
- The safety of the microorganism or enzyme has to be demonstrated.
- Microbial and enzyme additives have to be stable during storage and must be able to act in the complex environment of the gastrointestinal tract.
- The velocity of the detoxification reaction has to be high enough to enable the transformation of the mycotoxin molecule prior to its absorption from the gastrointestinal tract. Numerous studies have underlined this requirement and in some cases, for example, in case of the OTA-degrading strain

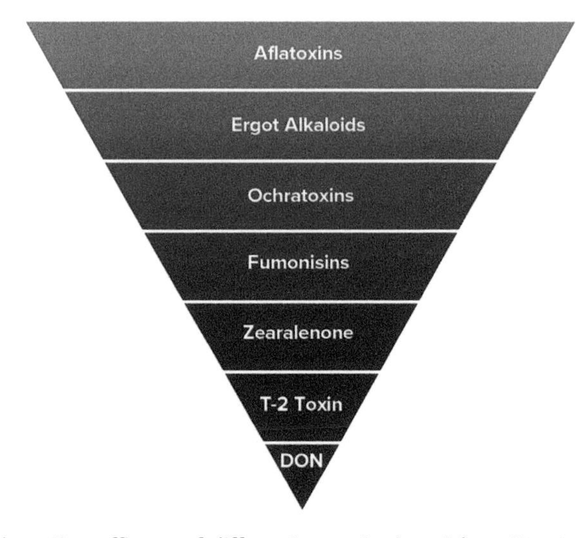

***Figure 6.2** – Adsorption efficacy of different mycotoxins. Adsorption is a suitable strategy for aflatoxins, ergot alkaloids and ochratoxins, but it is not an efficient method to counteract trichothecenes, fumonisins and zearalenone.*

Phenylobacterium immobile (Wegst and Lingens, 1983) the degradation velocity was too low for an effective application in animal nutrition (Schatzmayr *et al.*, 2006c).
- The efficacy of mycotoxin detoxifying microorganisms and enzymes must be demonstrated in feeding trials with target animals.

If these requirements are met, biotransformation represents an innovative way to counteract mycotoxins in animal feed.

Many experiments on trichothecene biotransformation that have been performed during the past 30 years focused on rumen fluid and intestinal contents. Rumen fluid was investigated due to the well-known higher resistance of ruminants to the negative effects of trichothecenes in comparison with monogastric animals. In 1983, Yoshizawa *et al.* reported the degradation of DON to de-epoxy-deoxynivalenol (DOM-1) in rats. In the following years, many researchers were able to show the degradation of DON to DOM-1 by ruminal or gut microflora *in vitro* (He *et al.*, 1992; King *et al.*, 1984; Kollarczik *et al.*, 1994; Swanson *et al.*, 1987). However, no pure culture of a DON degrading strain could be obtained until Schatzmayr *et al.* (2006a) succeeded in isolating the bacterial strain BBSH 797 capable of converting DON to DOM-1 (Figure 6.3).

The toxicity of DOM-1 was tested using a chicken lymphocyte proliferation assay (Schatzmayr *et al.*, 2006b). Application of DON at a concentration of 0.63 μg ml^{-1} inhibited lymphocyte proliferation completely, whereas application of a much higher concentration of DOM-1 (116 μg ml^{-1}) was necessary. Consequently, DOM-1 showed a dramatically reduced toxicity compared with DON. In line with this evidence, Pierron *et al.* (2016b) reported that – unlike DON – DOM-1 does not impair the proliferation, viability or barrier function of human intestinal cells. Furthermore, in intestinal explants obtained from pigs, DOM-1 does not induce histological alterations or inflammation and does not affect the global gene expression profile (Pierron, 2016b). Just recently, a study compared the effect of DON and DOM-1 on porcine intestinal epithelial cells by measuring six cytotoxicity parameters (Springler *et al.*, 2017). A lack of toxicity of DOM-1 was observed up to the highest tested concentration (100 μM). Another recent study investigated how DON and DOM-1 affect cell lines differing in tissue origin and species origin (trout gill cells, pig intestinal cells, mouse immune cells, human liver cells) (Mayer *et al.*, 2017). This is the first study to evaluate the effects of DOM-1 also in fish. Unlike DON, DOM-1 did not decrease the viabilities of the different cells. Only in liver cells, albumin secretion was also reduced by DOM-1, but a much higher concentration of DOM-1 (228 μmol L^{-1}) was needed compared with DON (0.9 μmol L^{-1}). Albumin is a blood serum protein synthesized in the liver and measured to observe liver damage.

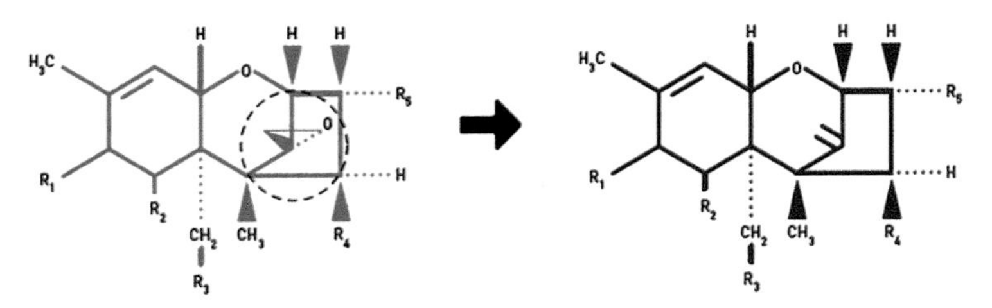

***Figure 6.3** – Basic molecular structure of trichothecenes before (left) and after (right) transformation by BBSH 797.*

BBSH 797 is the only microorganism that is available as a commercial product for the degradation of trichothecenes in animal feed (EFSA, 2009). Product development included the optimization of the fermentation procedure to facilitate a high growth rate of the organism, and the development of an encapsulation process to ensure a high stability of the product during storage and during the passage through the digestive tract.

Trichosporon mycotoxinivorans (MTV), a species of basidiomycete yeast capable of assimilating inulin and galacticol, was isolated from the hindgut of lower termites. MTV was found to biotransform both OTA and ZEN (Molnar *et al.*, 2004; Schatzmayr *et al.*, 2003, 2006b). Since *T. mycotoxinivorans* can be fermented, concentrated, freeze-dried and stabilized without losing its mycotoxin transforming abilities, its utilization as a feed additive for mycotoxin detoxification seemed feasible.

MTV detoxifies OTA by the cleavage of its amide bond resulting in the formation of phenylalanine and ochratoxin α (OTα) (Figure 6.4). A macrophage activation test showed that OTα is considerably less toxic than OTA (Schatzmayr *et al.*, 2006b). The growth of macrophages was decreased when OTA was applied at concentrations of 0.741 to 2.222 μg ml^{-1}. At concentrations above 6.667 μg OTA ml^{-1}, their growth was completely inhibited. On the other hand, concentrations up to 20 μg OTα ml^{-1} did not affect the growth of macrophages (Schatzmayr *et al.*, 2006b). These results were in accordance with those of other scientific studies where OTα was shown to be non-toxic or at least 500 times less toxic than OTA (Bruinink *et al.*, 1998; Chu *et al.*, 1972).

MTV converts ZEN to the nonestrogenic metabolite ZOM-1 (Figure 6.5) (Vekiru *et al.*, 2010). ZOM-1 did not induce a response that would indicate an estrogenic activity when tested in an E-screen assay, a commonly used system for evaluating the ability of chemicals to induce a hormonal response (Vekiru *et al.*, 2010).

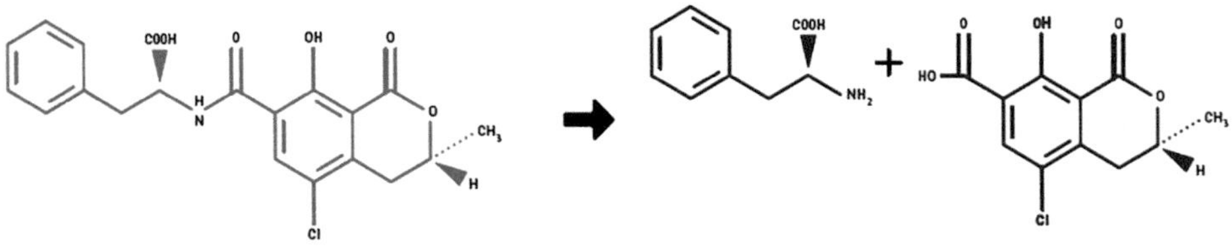

***Figure 6.4** – Ochratoxin A (left) is transformed into the less toxic metabolite ochratoxin α and phenylalanine (right) by the yeast MTV.*

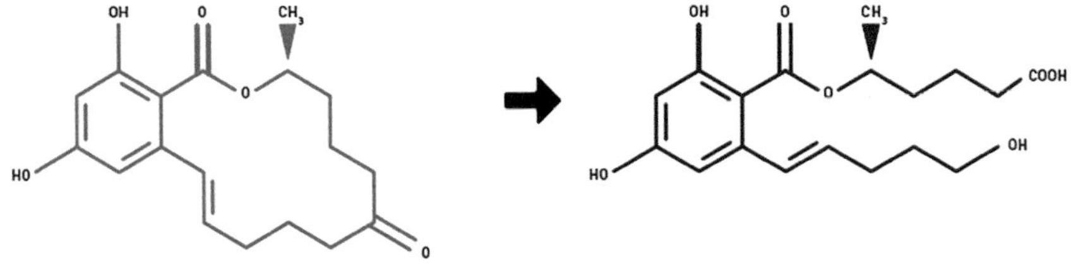

***Figure 6.5** – Zearalenone (left) is transformed into the non-estrogenic metabolite ZOM-1 (right).*

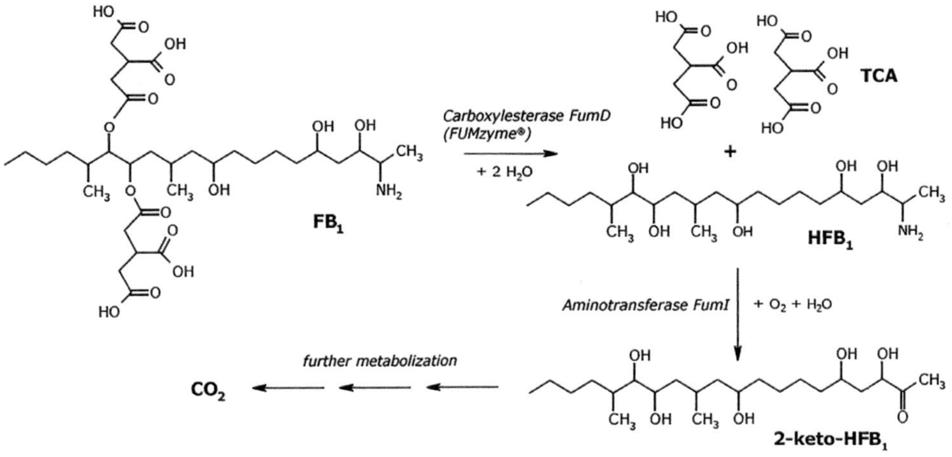

Figure 6.6 *– Degradation of fumonisin B_1 by Sphingopyxis MTA 144.*
Note: FB_1 = fumonisin B_1, HFB_1 = hydrolyzed FB_1, TCA = trichloroacetic acid.

The bacterial strain *Sphingopyxis* MTA 144 catalyzes the degradation of fumonisin B_1 (FB_1). The identification of its genes fumD and fumI that encode the enzymes responsible for this degradation provided a basis for the development of an enzymatic detoxification process for FB_1 in food and feed (Hartinger and Moll, 2011; Heinl *et al.*, 2010). The FUM carboxylesterase FumD converts FB_1 into hydrolyzed FB_1 (HFB_1) (Heinl *et al.*, 2010) and the aminotransferase FumI deaminates HFB_1 (Hartinger *et al.*, 2011; Heinl *et al.*, 2010) (Figure 6.6). Research demonstrated that HFB_1 did not cause intestinal or hepatic toxicity in a very sensitive pig model and disrupted sphingolipid metabolism only slightly (Grenier *et al.*, 2012). This finding indicates that enzymatic conversion of FB_1 into HFB_1 is a feasible strategy to counteract FB_1 exposure. Consequently, FumD was developed into the commercial product FUM*zyme*® that represents the first purified mycotoxin degrading enzyme applicable as a feed additive.

6.3 Acknowledgements

This text was originally published in *Guide to Mycotoxins* (Krska *et al.*, 2012) and written by Inês Rodrigues, Karin Naehrer and Christine Schwab, updated by Anneliese Mueller.

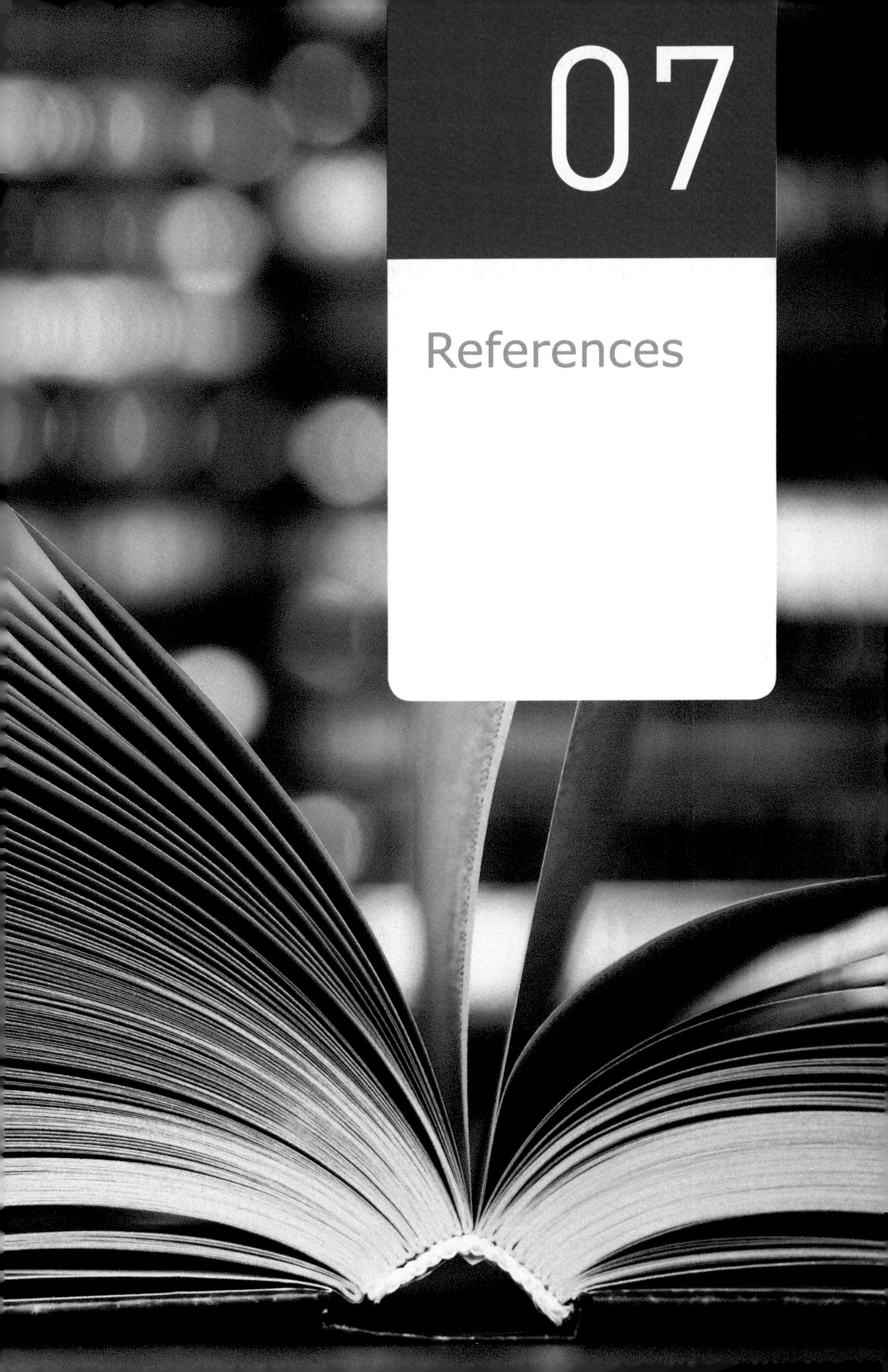

07

References

Abbas, A.K. and Lichtman, A.H. (2003). *Fondamenti di immunologia, funzioni ed alterazioni del sistema immunitario*. Padova, Piccin Nuova Libraria S.p.A.

Abdelhamid, A., Khalil, F. and Ragab, M. (1998). Problem of mycotoxins in fish production. *Egyptian Journal of Nutrition and Feeds* 1:63–71.

Adem, H.N., Tressel, R.-P., Pudel, F., Slawski, H. and Schulz, C. (2014). Rapeseed use in aquaculture. *Oilseeds and fats, Crops and Lipids* 21:1–9.

Afriyie-Gyawu, E., Wiles, M.C., Huebner, J.H., Richardson, B.M., Fickey, C. and Phillips, T.D. (2005). Prevention of zearalenone induced hyperestrogenism in prepubertal mice. *Journal of Toxicology and Environmental Health Part A* 12:353–368.

Agouz, H.M. and Anwer, W. (2011). Effect of Biogen® and Myco-Ad® on the growth performance of common carp (*Cyprinus carpid*) fed a mycotoxin contaminated aquafeed. *Journal of Fisheries and Aquatic Science* 6:334–345.

Alassane-Kpembi, I., Schatzmayr, G., Taranu, I., Marin, D., Puel, O. and Oswald, I.P. (2016). Mycotoxins co-contamination: Methodological aspects and biological relevance of combined toxicity studies. *Critical Reviews in Food Science and Nutrition* 57(16):3489–3507.

Alassane-Kpembi, I., Kolf-Clauw, M., Gauthier, T., Abrami, R., Abiola, F.A., Oswald, I.P. and Puel, O. (2013). New insights into mycotoxin mixtures: the toxicity of low doses of Type B trichothecenes on intestinal epithelial cells is synergistic. *Toxicology and Applied Pharmacology* 272:191–198.

Alberts, J.F., Engelbrecht, Y., Steyn, P.S., Holzapfel, W.H. and van Zyl, W.H. (2006). Biological degradation of aflatoxin B_1 by *Rhodococcus erythropolis* cultures. *International Journal of Food Microbiology* 109:121–126.

Ali, N., Sardjono, Yamashita, A. and Yoshizawa, T. (1998). Natural co-occurrence of aflatoxins and Fusarium mycotoxins (Fumonisins, Deoxynivalenol, Nivalenol and Zearalenone) in corn from Indonesia. *Food Additives & Contamination: Part A* 15:377–384.

Anater, A., Manyes, L., Meca, G., Ferrer, E., Luciano, F.B., Pimpão, C.T. and Font G. (2016). Mycotoxins and their consequences in aquaculture: A review. *Aquaculture* 451:1–10.

Antonissen, G., van Immerseel, F., Pasmans, F., Ducatelle, R., Haesebrouck, F., Timbermont, L., Verlinden, M., Janssens, G.P.J., Eeckhaut, V., Eeckhout, M., De Saeger, S., Hessenberger, S., Martel, A. and Croubels, S. (2014). The mycotoxin deoxynivalenol predisposes for the development of clostridium perfringens-induced necrotic enteritis in broiler chickens. *PLoS ONE* 9(9):e108775.

Arana, S., Alves, V.A.F., Sabino, M., Tabata, Y.A., Nonogaki, S., Zaidan-Dagli, M.L. and Hernandez-Blazquez, F.J. (2013). Immunohistochemical Evidence for Myofibroblast-like Cells Associated with Liver Injury Induced by Aflatoxin B_1 in Rainbow Trout (*Oncorhynchus mykiss*). *Journal of Comparative Pathology* 150:258–265.

Arukwe, A., Grotmol, T., Haugen, T.B., Knudsen, F.R. and Goksøyr, A. (1999). Fish model for assessing the in vivo estrogenic potency of the mycotoxin zearalenone and its metabolites. *Science of the Total Environment* 236:153–161.

Avantaggiato, G., Solfrizzo, M. and Visconti, A. (2005). Recent advances on the use of adsorbent materials for detoxification of Fusarium mycotoxins. *Food Additives and Contaminants* 22:379–388.

Awad, W.A., Ghareeb, K., Böhm, J. and Zentek, J. (2010). Decontamination and detoxification strategies for the Fusarium mycotoxin deoxynivalenol in animal feed and the effectiveness of microbial biodegradation. *Food Additives and Contaminants* 27(4):510–520.

Aziz, N.H. and Moussa, L.A.A. (2004). Reduction of fungi and mycotoxins formation in seeds by gamma-irradiation. *Journal of Food Safety* 24:109–127.

Bailey, G.S., Loveland, P.M., Pereira, A., Pierce, D., Hendricks, J.D. and Groopman, J.D. (1994). Quantitative carcinogenesis and dosimetry in rainbow trout for aflatoxin B_1 and aflatoxicol, two aflatoxins that form the same DNA adduct. *Mutation Research/ Environmental Mutagenesis and Related Subjects* 313:25–38.

Bakos, K., Kovács, R., Staszny, Á., Sipos, D.K., Urbányi, B., Müller, F., Csenki, Z. and Kovács, B. (2013). Developmental toxicity and estrogenic potency of zearalenone in zebrafish (Danio rerio). *Aquatic Toxicology* 136:13–21.

Baldwin, T.T., Riley, R.T., Zitomer, N.C., Voss, K.A., Coulombe Jr., R.A., Pestka, J.J., Williams, D.E. and Gelenn, A.E. (2011). The current state of mycotoxin biomarker development in humans and animals and the potential for application to plant systems. *World Mycotoxin Journal* 4:257–270.

Barbosa, T., Pereyra, C., Soleiro, C., Dias, E., Oliveira, A., Keller, K., Silva, P.P., Cavaglieri, L. and Rosa, C.A. (2013). Mycobiota and mycotoxins present in finished fish feeds from farms in the Rio de Janeiro State, Brazil. *International Aquatic Research* 5:3–11.

Battacone, G., Nudda, A. and Pulina, G. (2010). Effects of ochratoxin A on livestock production. *Toxins* 2:1796–1824.

Benbrook, C.M. (2005). *Breaking the Mold – Impacts of Organic and Conventional Farming Systems on Mycotoxins in Food and Livestock Feed*. The Organic Center.

Bennett, J.W. and Klich, M. (2003). Mycotoxins, *Clinical Microbiology Reviews* 16:497–498.

Berde, B. (1980). Ergot compounds: A synopsis. In: Goldstein, M., Lieberman, A., Calne, D.B. and Thorner, M.O. (eds) *Ergot Compounds and Brain Function: Neuroendocrine and Neuropsychiatric Aspects*. Raven Press, New York, NY, pp.4–23.

Berthiller, F., Crews, C., Dall'Asta, C., Saeger, S.D., Haesaert, G., Karlovsky, P., Oswald, I.P., Seefelder, W., Speijers, G. and Stroka, J. (2013). Masked mycotoxins: a review. *Molecular Nutrition & Food Research* 57:165–186.

Berthiller, F., Krska, R., Domigc, K.J., Kneifeld, W., Jugee, N., Schuhmacher, R. and Adam, G. (2011). Hydrolytic fate of deoxynivalenol-3-glucoside during digestion. *Toxicology Letters* 206:264–267.

Berthiller, F., Schuhmacher, R., Adam, G. and Krska, R. (2009). Formation, determination and significance of masked and other conjugated mycotoxins. *Analytical and Bioanalytical Chemistry* 395:1243–1252.

Biehl, M.L., Prelusky, D.B., Koritz, G.D., Hartin, K.E., Buck, W.B. and Trenholm, H.L. (1993). Biliary excretion and enterohepatic cycling of zearalenone in immature pigs. *Toxicology and Applied Pharmacology* 121:152–159.

Bintvihok, A., Ponpornpisit, A., Tangtrongpiros, J., Panichkriangkrai, W., Rattanapanee, R., Doi, K. and Kumagai, S. (2003). Aflatoxin contamination in shrimp feed and effects of aflatoxin addition to feed on shrimp production. *Journal of Food Protection* 66:882–885.

Bojana Kokic, I.C., Jovanka Levic, Anamarija Mandic, Jovana Matic and Dusica Ivanov. (2009). Screening of mycotoxins in animal feed from the region of Vojvodina. *Zbornik Matice srpske za prirodne nauke* (*Matica Srpska Journal for Natural Sciences*) 117:87–96.

Boonyaratpalin, M., Supamattaya, K., Verakunpiriya, V. and Suprasert, D. (2001). Effects of aflatoxin B_1 on growth performance, blood components, immune function and histopathological changes in black tiger shrimp (*Penaeus monodon* Fabricius). *Aquaculture Research* 32:388–398.

Boudra, H., Le Bars, P. and Le Bars, J. (1995). Thermostability of ochratoxin A in wheat under two moisture conditions. *Applied and Environmental Microbiology* 61:1156–1158.

Braun, M.S. and Wink, M. (2018). Exposure, occurrence, and chemistry of fumonisins and their cryptic derivatives. *Comprehensive Reviews in Food Science and Food Safety* 17:769–791. Doi: 10.1111/1541-4337.12334.

Broekaert, N., Devreese, M., van Bergen, T., Schauvliege, S., De Boevre, M., De Saeger, S., Vanhaecke, L., Berthiller, F., Michlmayr, H., Malachová, A., Adam, G., Vermeulen, A., Croubels, S. (2016). *In vivo* contribution of deoxynivalenol-3-β-D-glucoside to deoxynivalenol exposure in broiler chickens and pigs: oral bioavailability, hydrolysis and toxicokinetics. *Archives of Toxicology* 91:699–712.

Bruinink, A., Rasonyi, T. and Sidler, C. (1998). Differences in neurotoxic effects of ochratoxin A, ochracin and ochratoxin-α in vitro. *Natural Toxins* 6:173–177.

Bullerman, L.B. and Bianchini, A. (2007) Stability of mycotoxins during food processing. *International Journal of Food Microbiology* 119:140–146.

Bundit, O., Kanghae, H., Phromkunthong, W. and Supamattaya, K. (2006). Effects of mycotoxin T-2 and zearalenone on histopathological changes in black tiger shrimp (*Penaeus monodon* Fabricius). *Journal of Science and Technology* 28(5):937–949.

Bürk, G., Höbel, W. and Richt, A. (2006) Ergot alkaloids in cereal products. Results from the Bavarian Health and Food Safety Authority. *Molecular Nutrition and Food Research* 50:437–442.

Busby, W.F. and Wogan, G.N. (1984). Aflatoxins. In: Searle, C.E.E. (ed.), *Chemical Carcinogenesis*. American Chemical Society, Washington, DC. pp.945–1136.

CAC (Codex Alimentarius Commission). (2003a). Joint FAO/WHO Food Standards Programme Codex Alimentarius Commission, Twenty-sixth Session, Rome, Italy, and Report of the 35th Session of the Codex Committee on Food Additives and Contaminants. Arusha, Tanzania. CX 4/30.2 CL 2003/13-FAC.

CAC (Codex Alimentarius Commission). (2003b). Discussion on deoxynivalenol, CX/FAC03/35, Joint FAO/WHO Food Standards Programme. Arusha, Tanzania.

Caloni, F. and Cortinovis, C. (2010). Effects of fusariotoxins in the equine species. *The Veterinary Journal* 186:157–161.

Carlson, D.B. and Williams, D.E., Spitsbergen, J.M., Ross, P.F., Bacon, C.W., Meredith, F.I. and Riley, R.T. (2001). Fumonisin B_1 Promotes Aflatoxin B_1 and N-Methyl-N′-nitro-nitrosoguanidine-Initiated Liver Tumors in Rainbow Trout. *Toxicology and Applied Pharmacology* 172:29–36.

CAST (2003). Mycotoxins: risks in plant, animal, and human systems (Richard, J.L. and Payne, G.A. (eds)). Council for Agricultural Science and Technology Task Force report No. 139, Ames, Iowa, USA.

Centoducati, G., Santacroce, M.P., Lestingi, A., Casalino, E. and Crescenzo, G. (2010). Characterization of the cellular damage induced by Aflatoxin B_1 in sea bream (*Sparus aurata* Linnaeus, 1758) hepatocytes. *Italian Journal of Animal Science* 8: Proceedings of the 18th ASPA Congress, Palermo, June 9–12, 2009.

Chávez-Sánchez, M.C., Martínez Palacios, C.A. and Osorio, M.I. (1994). Pathological effects of feeding young *Oreochromis niloticus* diets supplemented with different levels of aflatoxin B_1. *Aquaculture* 127:49–60.

Cheli, F., Campagnoli, A., Pinotti, L., Fusi, E. and Dell`Orto, V. (2009). Sampling feed for mycotoxins: acquiring knowledge from food. *Italian Journal of Animal Science* 8:5–22.

Cheli, F., Pinotti, L., Rossi, L. and Dell'Orto, V. (2013). Effect of milling procedures on mycotoxin distribution in wheat fractions: A review. *LWT – Food Science and Technology* 54:307–314.

Chen, H., Hu, J., Yang, J., Wang, Y., Xu, H., Jiang, Q., Gong, Y., Gu, Y. and Song, H. (2010). Generation of a fluorescent transgenic zebrafish for detection of environmental estrogens. *Aquatic Toxicology* 96(1):53–61.

Chu, F.S., Noh, I. and Chang, C.C. (1972). Structural Requirements for Ochratoxin Intoxication. *Life Sciences* 11:503–508.

Ciegler, A., Lillehoj, B., Peterson, R.E. and Hall, H.H. (1966) Microbial detoxification of aflatoxin. *Applied Microbiology* 39:139–143.

Claudino-Silva, S.C., Lala, B., Mora, N.H.A.P., Schamber, C.R., Nascimento, C.S., Pereira, V.V., Hedler, D.L. and Gasparino, E. (2018). Challenge with fumonisins B_1 and B_2 changes IGF-1 and GHR mRNA expression in liver of Nile tilapia fingerlings. *World Mycotoxin Journal* 11:237–245.

Cooper M.D., Alder M A. (2006). The evolution of adaptive immune systems. *Cell* 124:815–822.

Creppy, E.E. (2002). Update of survey, regulation and toxic effects of mycotoxins in Europe. *Toxicology Letters* 127:19–28.

D'Mello, J.P.F. and Macdonald, A.M.C. (1997). Mycotoxins. *Animal Feed Science and Technology* 69:155–166.

Dabrowski, K., Lee, K.J., Rinchard, J., Ciereszko, A., Blom, J.H. and Ottobre, J. (2001). Gossypol isomers bind specifically to blood plasma proteins and spermatozoa of rainbow trout fed diets containing cottonseed meal. *Biochimica et Biophysica Acta General Subjects* 1525:37–42.

Daković, A., Tomasevic-Canovic, M., Dondur, V., Rottinghaus, G.E., Medakovic, V. and Zaric, S. (2005). Adsorption of mycotoxins by organozeolites. *Colloids and Surfaces B: Biointerfaces* 46:20–25.

Daković, A., Tomasevic-Canovic, M., Rottinghaus, G.E., Matijasevic, S. and Sekulic, Z. (2007). Fumonisin B_1 adsorption to octadecyldimenthylbenzyl ammonium-modified clinoptilolite-rich zeolitic tuff. *Microporous and Mesoporous Materials* 105:284–290.

Dall'Asta, C., Lindner, J.D., Galaverna, G., Dossena, A., Neviani, E. and Marchelli, R. (2008). The occurrence of ochratoxin A in blue cheese. *Food and Chemical Toxicology* 106:729–734.

Davis, D.A. and Sookying, D. (2009). Strategies for reducing and/or replacing fishmeal in production diets for the Pacific white shrimp, *Litopenaeus vannamei*. In: Browdy C.L. and Jory D.E. (eds), *The Rising Tide, Proceedings of the Special Session on Sustainable Shrimp Farming.*, Baton Rouge, USA, World Aquaculture Society. *World Aquaculture,* pp. 108–114.

De Boevre, M., Di Mavungu, D.J., Landschoot, S., Audenaert, K., Eeckhout, M., Maene, P., Haesaert, G. and De Saeger, S. (2012). Natural occurrence of mycotoxins and their masked forms in food and feed products. *World Mycotoxin Journal* 5:207–219.

Deng, Y., Barrientos Velazquez, A.L., Billes, F. and Dixon, J.B. (2010). Bonding mechanisms between aflatoxin B_1 and smectite. *Applied Clay Science* 50:92–98.

Deng, S.X., Tian, L.X., Liu, F.J., Jin, S.J., Liang, G.Y., Yang, H.J., Du, Z.Y. and Liu, Y.J. (2010). Toxic effects and residue of aflatoxin B_1 in tilapia (*Oreochromis niloticus* × *O. aureus*) during long-term dietary exposure. *Aquaculture* 307:233–240.

Desai, K., Sullards, M.C., Allegood, J., Wang, E., Schmelz, E.M., Hartl, M., Humpf, H.U., Liotta, D.C., Peng. Q., and Merrill, A.H. (2002). Fumonisins and fumonisin analogs as inhibitors of ceramide synthase and inducers of apoptosis. *Biochimia et Biophysia Acta* 1585:188–192.

Dill-Macky, R. and Jones, R.K. (2000). The effect of previous crop residues and tillage on Fusarium head blight of wheat. *Plant Disease* 48(1):71–76.

Dirican, S. (2015). A review of effects of aflatoxins in aquaculture. *Applied Research Journal* 1:192–196.

Döll, S., Dänicke, S. and Valenta, H. (2008). Residues of deoxynivalenol (DON) in pig tissue after feeding mash or pellet diets containing low concentrations. *Molecular Nutrition and Food Research* 52:727–374.

Döll, S., Baardsen, G., Koppe, W., Stubhaug, I. and Dänicke, S. (2010). Effects of increasing concentrations of the mycotoxins deoxynivalenol, zearalenone or ochratoxin A in diets for Atlantic salmon (*Salmo salar*) on growth performance and health. The 14th International Symposium on Fish Nutrition and Feeding, Qingdao, China, pp.120.

Doster, R.C., Sinnhuber, R.O. and Wales, J.H. (1972). Acute intraperitoneal toxicity of ochratoxins A and B in rainbow trout (*Salmo gairdneri*). *Food and Cosmetics Toxicology* 10:85–92.

Duncan, K.E. and Howard, R.J. (2010). Biology of Maize Kernel Infection by *Fusarium verticillioides*. *Molecular Plant-Microbe Interactions* 23:6–16. Doi:10.1094/MPMI -23-1-0006.

Eaton, D.L., Beima, K.M., Bammler, T.K., Riley, R.T. and Voss, K.A. (2010). Hepatotoxic mycotoxins. In: Roth, R.A., Ganey, P.E. (eds), *Comprehensive Toxicology*, second ed. Vol. 10. Elsevier, New York, pp.527–569.

EC (2002). Directive 2002/32/EC of the European Parliament and of the Council of 7 May 2002 on undesirable substances in animal feed – Council statement.

EC (2006). Commission Recommendation No. 2006/576 of 17 August 2006 on the presence of deoxynivalenol, zearalenone, ochratoxin A, T-2 and HT-2 and fumonisins in products intended for animal feeding. *Official Journal of the European Union*: 7–9.

EFSA (European Food Safety Authority). (2004a). Opinion of the scientific panel on contaminants in food chain on a request from the commission related to ochratoxin A (OTA) as undesirable substance in animal feed. *EFSA Journal* 101:1–36.

EFSA (European Food Safety Authority). (2004b). Opinion of the scientific panel on contaminants in the food chain [CONTAM] related to Aflatoxin B_1 as undesirable substance in animal feed. *EFSA Journal* 2:39.

EFSA (European Food Safety Authority). (2004c). Opinion of the scientific panel on contaminants in the food chain on a request from the commission related to deoxynivalenol (DON) as undesirable substance in animal feed. *EFSA Journal* 73:1–42.

EFSA (European Food Safety Authority). (2004d). Opinion of the scientific panel on contaminants in the food chain on a request from the commission related to zearalenone as undesirable substance in animal feed. *EFSA Journal* 89:1–41.

EFSA (European Food Safety Authority). (2005a). Opinion of the scientific panel on contaminants in food chain on a request from the commission related to ergot as undesirable substance in animal feed. *EFSA Journal* 225:1–27.

EFSA (European Food Safety Authority). (2005b). Opinion of the scientific panel on contaminants in food chain on a request from the commission related to fumonisins as undesirable substances in animal feed. *EFSA Journal* 235:1–32.

EFSA (European Food Safety Authority). (2009). Review of mycotoxin-detoxifying agents used as feed additives: mode of action, efficacy and feed/food safety. External scientific report, pp. 1–192.

EFSA (European Food Safety Authority). (2011a). Survey on ergot alkaloids in cereals intended for human consumption and animal feeding. External scientific report, pp. 1–112.

EFSA (European Food Safety Authority). (2011b). Scientific opinion on the risks for animal and public health related to the presence of T-2 and HT-2 toxin in food and feed. *EFSA Journal* 9(12):2481.

EFSA (European Food Safety Authority). (2013). Deoxynivalenol in food and feed: occurrence and exposure. *EFSA Journal* 11:3379.

EFSA (European Food Safety Authority). (2018). Risks to human and animal health related to the presence of moniliformin in food and feed. *EFSA Journal* 16:5082.

El-Banna, R., Teleb, H.M., Hadi, M.M. and Fakhry, F.M. (1992). Performance and tissue residue of tilapias fed dietary aflatoxin. *Veterinary Medicine Journal (Giza)* 40:17–23.

El-Sayed, Y.S. and Khalil, R.H. (2009). Toxicity, biochemical effects and residue of aflatoxin B_1 in marine water-reared sea bass (*Dicentrarchus labrax* L.). *Food and Chemical Toxicology* 47:1606–1609.

El-Sayed, Y.S., Khalil, R.H. and Saad, T.T. (2009). Acute toxicity of ochratoxin-A in marine water-reared sea bass (*Dicentrarchus labrax* L.). *Chemosphere* 75:878–882.

Escriva, L., Font, G. and Manyes, L. (2015). In vivo toxicity studies of fusarium mycotoxins in the last decade: A review. *Food and Chemical Toxicology* 78:185–206.

Fajardo, J.E., Dexter, J.E., Roscoe, M.M. and Nowicki, T.W. (1995). Retention of ergot alkaloids in wheat during processing. *Cereal Chemistry* 72:291–298.

FAO (2004). Worldwide regulations for mycotoxins in food and feed in 2003. Food and Agriculture Organization of the United Nations, Rome.

Farmer, G.J., Ashfield, D. and Goff, T.R. (1983). A feeding guide for juvenile Atlantic salmon. *Canadian Manuscript Report of Fisheries and Aquatic Sciences* 1718:1–13.

Fegan, D.F. and Spring, P. (2007). Recognizing the reality of the aquaculture mycotoxin problem: searching for a common and effective solution. Nutritional Biotechnology in the Feed and Food Industries: Proceedings of Alltech's 23rd Annual Symposium, The New Energy Crisis: Food, Feed or Fuel?, Lexington, Kentucky, USA, 20–23 May 2007. Alltech UK, Stamford, pp.343–354.

Fink-Gremmels, J. and Malekinejad, H. (2007). Clinical effects and biochemical mechanisms associated with exposure to the mycoestrogen zearalenone. *Animal Feed Science and Technology* 137:326–341.

Flieger, M., Wurst, M. and Shelby, R. (1997). Ergot alkaloids - Sources, structures and analytical methods. *Folia Microbiologica* 42:3–29.

Foey, A., Picchietti, S. (2014). Immune defense of teleost fish. In: *Aquaculture nutrition: Gut health, Probiotics and Prebiotics*, first edition.

Fowler, L.G. (1980). Substitution of soybean and cottonseed products for fish meal in diets fed to Chinook and Coho salmon. *Progressive Fish-Culturist* 42:87–91.

Frederickson, D.E., Mantle, P.G. and de Milliano. W.A.J. (1991). Claviceps africana sp. nov., the distinctive ergot pathogen of sorghum in Africa. *Mycological Research* 95:1101–1107.

Gallagher, E.P. and Eaton, D.L. (1995). In vitro biotransformation of Aflatoxin B_1 (AFB_1) in channel catfish liver. *Toxicology and Applied Pharmacology* 132:82–90.

García-Morales, M.-H., Pérez-Velázquez, M., González-Felix, M.L., Burgos-Hernández, A., Cortez-Rocha, M.-O., Bringas-Alvarado, L. and Ezquerra-Brauer, J.-M. (2013). Effects of fumonisin B_1-containing feed on the muscle proteins and ice-storage life of white shrimp (*Litopenaeus vannamei*). *Journal of Aquatic Food Product Technology* 24:340–353.

García, E.C. (2013). Effects of fumonisin B_1 on performance of juvenile Baltic salmon (*Salmo salar*). Masters thesis, University of Jyväskylä, Faculty of Science.

Gatlin, D., Barrows, F., Bellis, D., Brown, P., Campen, J., Dabrowski, K., Gaylord, T.G., Hardy, R.W., Herman, E.M., Hu, G., Krogdahl, A., Nelson, R., Overturf, K.E., Rust, M., Sealey, W., Skonberg, D., Souza, E.J., Stone, D. and Wilson, R.F. (2007). Expanding the utilization of sustainable plant products in aquafeeds: A review. *Aquaculture Research* 38:551–579.

Giovati, L., Magliani, W., Ciociola, W., Santinoli, C., Conti, S. and Polonelli, L. (2015). AFM1 in milk: physical, biological, and prophylactic methods to mitigate contamination. *Toxins* 7:4330–4349. Doi:10.3390/toxins7104330.

Goel, S., Lenz, S.D., Lumlertdacha, S., L., R.T., Shelby, R.A., Li, M., Riley, R.T. and Kemppainen, B.W. (1994). Sphingolipid levels in catfish consuming *Fusarium moniliforme* corn culture material containing fumonisins. *Aquatic Toxicology* 30:285–294.

Golli-Bennour, E.E., Kouidhi, B., Bouslimi, A., Abid-Essefi, S., Hassen, W. and Bacha, H. (2010). Cytotoxicity and genotoxicity induced by aflatoxin B_1, ochratoxin A, and their combination in cultured Vero cells. *Journal of Biochemical and Molecular Toxicology* 24:42–50.

Gonçalves, R.A., Schatzmayr, D., Hofstetter, U. and Santos, G.A. (2017). Occurrence of mycotoxins in aquaculture: preliminary overview of Asian and European plant ingredients and finished feeds. *World Mycotoxin Journal* 10:183–194.

Gonçalves, R.A., Do Cam, T., Tri, N.N., Santos, G.A., Encarnação, P. and Hung, L.T. (2018a). Aflatoxin B_1 (AFB_1) reduces growth performance, physiological response, and disease resistance in Tra catfish (*Pangasius hypophthalmus*). *Aquaculture International* 26:921–936.

Gonçalves, R.A., Hofstetter, U., Schatzmayr, D. and Jenkins, T. (2018b). Mycotoxins in Southeast Asian aquaculture: plant-based meals and finished feeds. *World Mycotoxin Journal* 11:265–275.

Gonçalves, R.A., Navarro-Guillén, C., Gilannejad, N., Dias, J., Schatzmayr, D., Bichl, G., Czabany, T., Moyano, F.J., Rema, P., Yúfera, M., Mackenzie, S. and Martínez-Rodríguez, G. (2018c). Impact of deoxynivalenol on rainbow trout: growth performance, digestibility, key gene expression regulation and metabolism. *Aquaculture* 490:362–372.

Gonçalves, R.A., Naehrer, K. and Santos, G.A. (2018d). Occurrence of mycotoxins in commercial aquafeeds in Asia and Europe: a real risk to aquaculture? *Reviews in Aquaculture* 10:263–280.

Gonçalves, R.A., Menanteau-Ledouble, S., Schöller, M., Eder, A., Schmidt-Posthaus, H., Mackenzie, S. and El-Matbouli, M. (2018e). Effects of deoxynivalenol exposure time and contamination levels on rainbow trout. *Journal of the World Aquaculture Society* in press.

Gonçalves, R.A., Dias, J., Serradeiro, R. and Schatzmayr, D. (in press-a). Impact of graded dietary fumonisin contamination levels on the growth performance and selected health indices of turbot (*Psetta maxima*). *Aquaculture International.*

Gonçalves, R.A., Schatzmayr, D., Albalat, A. and Mackenzie, S. (in press-b). Mycotoxins in aquaculture feed and food. *Reviews in Aquaculture.*

Goyarts, D., Dänicke, S., Valenta, H., and Ueberschär, K.H. (2007). Carry-over of Fusarium toxins (deoxynivalenol and zearalenone) from naturally contaminated wheat to pigs. *Food Additives and Contaminants* 24:369–380.

Grant, P.G. and Phillips, T.D. (1998). Isothermal adsorption of aflatoxin B_1 on HSCAS clay. *Journal of Agriculture and Food Chemistry* 46:599–605.

Gratz, S. (2007). Aflatoxin binding by probiotics: experimental studies on intestinal aflatoxin transport, metabolism and toxicity. Kuopio University Publications, Department of Medical Sciences 404. 85 pp.

Grenier, B. and Oswald. I.P. (2011). Mycotoxin co-contamination of food and feed: meta-analysis of publications describing toxicological interactions. *World Mycotoxin Journal* 4:285–313.

Grenier, B. and Applegate, T.J. (2013). Modulation of intestinal functions following mycotoxin ingestion: Meta-analysis of published experiments in animals. *Toxins* 5:396–430.

Grenier, B., Bracarense, A.P.F.L., Schwartz, H.E., Trumel, C., Cossalter, A.M., Schatzmayr, G., Kolf-Clauw, M., Moll. W.D. and Oswald, I.P. (2012). The low intestinal and hepatic toxicity of hydrolyzed fumonisin B_1 correlates with its inability to alter the metabolism of sphingolipids. *Biochemical Pharmacology* 83:1465–1473.

Gruber-Dorninger, C., Novak, B., Nagl, V. and Berthiller, F. (2017). Emerging mycotoxins: Beyond traditionally determined food. Contaminants. *Journal of Agriculture and Food Chemistry* 6:7052–7070.

Guillamont, E.M., Lino, C.M., Baeta, M.L., Pena, A.S., Silveira, M.I.N. and Vinuesa, J.M. (2005). A comparative study of extraction apparatus in HPLC analysis of ochratoxin A in muscle. *Analytical and Bioanalytical Chemistry* 383:570–575.

Hafner, M., Sulyok. M., Schuhmacher. R., Crews, C. and Krska, R. (2008). Stability and epimerization behavior of ergot alkaloids in various solvents. *World Mycotoxin Journal* 1:67–78.

Halver, J.E. (1969). Aflatoxicosis and trout hepatoma. In: Goldblatt, L.A.E. (ed.), *Aflatoxin: Scientific Background, Control, and Implications*. Academic Press, New York, pp.265–306.

Hart, S.D., Bharadwaj, A.S. and Brown, P.B. (2010). Soybean lectins and trypsin inhibitors, but not oligosaccharides or the interactions of factors, impact weight gain of rainbow trout (*Oncorhynchus mykiss*). *Aquaculture* 306:310–314.

Hartinger D. and Moll, W.D. (2011). Fumonisin elimination and prospects for detoxification by enzymatic transformation. *World Mycotoxin Journal* 4:271–283.

Hartinger D., Schwartz H., Hametner C., Schatzmayr G., Haltrich D. and Moll, W.D. (2011). Enzyme characteristics of aminotransferase FUMI of Sphingopyxis sp. MTA 144 for deamination of hydrolyzed fumonisin B_1. *Applied Microbiology and Biotechnology* 91:757–768.

Harvey, R.B., Edrington, T.S. and Kubena, L.F. (1995). Influence of aflatoxin and fumonisin B_1 containing culture material on growing barrows. *American Journal of Veterinary Research* 56:1668–1672.

Hasan, M.R. (2007). Economics of aquaculture feeding practices in selected Asian countries. FAO Fisheries and Aquaculture Technical Paper No. 505.

Haschek, W.M., Gumprecht, L.A., Smith, G., Tumbleson, M.E. and Constable, P.D. (2001). Fumonisin toxicosis in swine: an overview of porcine pulmonary edema and current perspectives. *Environmental Health Perspectives* 109(Suppl. 2):251–257.

Hassan, A.M., Kenawy, A.M., Abbas, W.T. and Abdel-Wahhab, M.A. (2010). Prevention of cytogenetic, histochemical and biochemical alterations in *Oreochromis niloticus* by dietary supplement of sorbent materials. *Ecotoxicology and Environmental Safety* 73:1890–1895.

Hatai, K., Kubota, S.S., Kida N. and Udagawa, S.-I. (1986). *Fusarium oxysporum* in red sea bream (*Pagrus* spp.). *Journal of Wildlife Diseases* 22:570–571.

He, P., Young, L.G. and Forsberg, C. (1992). Microbial transformation of deoxynivalenol (vomitoxin). *Applied and Environmental Microbiology* 58:3857–3863.

He, C.H., Fan, Y.H., Wang, Y., Huang, C.Y., Wang, X.C. and Zhang, H.B. (2010). The individual and combined effects of deoxynivalenol and Aflatoxin B(1) on primary hepatocytes of Cyprinus carpio. *International Journal of Molecular Sciences* 11:3760–3768.

Heinl, S., Hartinger. D., Thamhesl, M., Veriku, W., Krska, R., Schatzmayr, G., Moll. W.D. and Grabherr, R. (2010). Degradation of fumonisin B_1 by the consecutive action of two bacterial enzymes. *Journal of Biotechnology* 145:120–129.

Hendricks, J.D. (1994). Carcinogenicity of aflatoxins in nonmammalian organisms. In: Eaton, D.L. and Groopman, J.D. (eds) *Toxicology of Aflatoxins: Human Health, Veterinary, and Agricultural Significance*. Academic Press, San Diego.

Hertrampf, J.W. and Piedad-Pascual, F. (2000a). Cotton-seed meal. *Handbook on Ingredients for Aquaculture Feeds*. Springer Netherlands, Dordrecht, pp.531–542.

Hertrampf, J.W. and Piedad-Pascual, F. (2000b). Maize products. *Handbook on Ingredients for Aquaculture Feeds*. Springer Netherlands, Dordrecht, pp.262–280.

Hertrampf, J.W. and Piedad-Pascual, F. (2000c). Rice by-products. *Handbook on Ingredients for Aquaculture Feeds*. Springer Netherlands, Dordrecht, pp.531–542.

Hertrampf, J.W. and Piedad-Pascual, F. (2000d). Shrimp meal. *Handbook on Ingredients for Aquaculture Feeds*. Springer Netherlands, Dordrecht, pp.364–371.

Hertrampf, J.W. and Piedad-Pascual, F. (2000e). Vegetable oil meals. *Handbook on Ingredients for Aquaculture Feeds*. Springer Netherlands, Dordrecht, pp.531–542.

Hertrampf, J.W. and Piedad-Pascual, F. (2000f). Wheat and wheat by-products. *Handbook on Ingredients for Aquaculture Feeds*. Springer Netherlands, Dordrecht, pp.531–542.

Hooft, J.M., Elmor A., Ibraheem, E.H., Encarnaçao, P. and Bureau, D.P. (2011). Rainbow trout (*Oncorhynchus mykiss*) is extremely sensitive to the feed-borne Fusarium mycotoxin deoxynivalenol (DON). *Aquaculture* 311:224–232.

Hooft, J.M. and Bureau, D.P. (2017). Evaluation of the efficacy of a commercial feed additive against the adverse effects of feed-borne deoxynivalenol (DON) on the performance of rainbow trout (*Oncorhynchus mykiss*). *Aquaculture* 473:237–245.

Huang, Y., Han, D., Xiao, X., Zhu, X., Yang, Y., Jin, J., Chen, Y. and Xie, S. (2014). Effect of dietary aflatoxin B_1 on growth, fecundity and tissue accumulation in gibel carp during the stage of gonad development. *Aquaculture* 428:236–242.

Humpf, H.U. and Voss, K.A. (2004). Effects of thermal food processing on the chemical structure and toxicity of fumonisin mycotoxins. Mold and Nutrition Food Research 48: 255–269.

Hussein, S.H. and Brassel, J.M. (2001). Toxicity, metabolism, and impact of mycotoxins on humans and animals. *Toxicology* 167:101–134.

Huwig, A., Freimund, S., Kappeli, O. and Dutler, H. (2001). Mycotoxin detoxication of animal feed by different adsorbents. *Toxicology Letters* 122: 179–188.

Iheshiulor, O.O.M., Esonu, B.O., Chuwuka, O.K., Omede, A.A., Okoli, I.C. and Ogbuewu, I.P. (2011). Effects of mycotoxins in animal nutrition: A review. *Asian Journal of Animal Sciences* 5:19–33. Doi: 10.3923/ajas.2011.19.33.

International Agency for Research on Cancer (IARC). (1993). Aflatoxins in some naturally occurring substances: Food items and constituents, heterocyclic aromatic amines, and mycotoxins. *IARC Monographs on the Evaluation of Carcinogenic Risk of Chemicals to Humans* 56:245–395.

International Agency for Research on Cancer (IARC). (2002). Aflatoxins in traditional herbal medicines, some mycotoxins, naphthalene and styrene. *IARC Monographs on the Evaluation of Carcinogenic Risks to Humans* 82:171–366.

International Agency for Research on Cancer (IARC). (2012). Chemical agents and related occupations. *IARC Monographs: A Review of Human Carcinogens* 100:225–248.

Jantrarotai, W. and Lovell, R.T. (1990). Subchronic toxicity of dietary aflatoxin B_1 to channel catfish. *Journal of Aquatic Animal Health* 2:248–254.

Jouany, J.P. (2007). Methods for preventing, decontaminating and minimizing the toxicity of mycotoxins in feeds. *Animal Feed Science and Technology* 137:342–362.

Kabak, B. (2009). The fate of mycotoxins during thermal food processing. *Journal of Science and Food Agriculture* 89:549–554.

Kannewischer, I., Tenorio, A.M.G., White, G.N. and Dixon, J.B. (2006). Smectite clays as adsorbents of aflatoxin B_1: initial steps. *Clay Science* 12(Supplement 2):199–204.

Khaneghah, A.M., Martins, L.M., von Hertwig, A.M., Bertoldo, R. and Sant'Ana, A.S. (2018). Deoxynivalenol and its masked forms: Characteristics, incidence, control and fate during wheat and wheat based products processing. A review. *Trends in Food Science and Technology* 71:13–24.

Khatibi, P.A., McMaster, N.J., Musser, R. and Schmale, D.G. (2014). Survey of mycotoxins in corn distillers' dried grains with solubles from seventy-eight ethanol plants in twelve states in the U.S. in 2011. *Toxins* 6:1155–1168.

King, R.R., McQueen, R.E., Levesque, D. and Greenhalgh, R. (1984). Transformation of deoxynivalenol (vomitoxin) by rumen microorganisms. *Journal of Agricultural and Food Chemistry* 32:1181–1183.

Kollarczik, B., Gareis, M. and Hanelt, M. (1994). In vitro transformation of the Fusarium mycotoxins deoxynivalenol and zearalenone by the normal gut microflora of pigs. *Natural Toxins* 2:105–110.

Kottapalli, B., Wolf-Hall, C.E., Schwarz, P., Schwarz, J. and Gillespie, J. (2003). Evaluation of hot water and electron beam irradiation for reducing Fusarium infection in malting barley. *Journal of Food Protection* 66:1241–1246.

Kren, V. and Cvak, L. (1999). Ergot the genus Claviceps. In: *Ergot alkaloids in animal feed*. Hardwood Academic Publishers. The Netherlands.

Krogdahl, A., Penn. M., Thorsen, J., Refstie, S. and Bakke, A.M. (2010). Important antinutrients in plant feedstuffs for aquaculture: an update on recent findings regarding responses in salmonids. *Aquaculture Research* 41:333–344.

Krska, R. and Crews, C. (2008). Significance, chemistry and determination of ergot alkaloids: A review. *Food Additives and Contaminants* 25:722–731.

Krska, R., Naehrer, K., Richard J.L., Rodrigues, I., Schuhmaber, R., Slate A.B. and Whitaker, T.B. (2012). *Guide to Mycotoxins*. Special Edition World Nutrition Forum 2012, Erber AG, Austria.

Kuiper-Goodman, T. (2004). Risk assessment and risk management of mycotoxins in food. In: Magan, N. and Olsen, M. (eds). *Mycotoxins in Food. Detection and Control*. CRC Press, Woodhead Publishing Limited, England, pp.3–31.

Kumar, V., Roy, S., Barman, D., Kumar, A., Paul, L. and Meetei, W.A. (2013). Importance of mycotoxins in aquaculture feeds. *Journal of Aquatic Animal Health* 18:25–29.

Lanza, G.M., Washburn, K.W. and Wyatt, R.D. (1980). Variation with age in response of broilers to aflatoxin. *Poultry Science* 59:282–288.

Leeman, W.R., Van den Berg, K.J. and Houben, G.F. (2007). Transfer of chemicals from feed to animal products: the use of transfer factors in risk assessment. *Food Additives & Contaminants* 24:1–13.

Leeson, S. and Summers, J.D. (1991). *Commercial poultry nutrition*. Department of Animal and Poultry Science. University of Guelph, Ontario, Canada. International book distributing Co.

Leeson, S., Diaz, G.J. and Summers, J.D. (1995). *Poultry Metabolic Disorders and Mycotoxins*. University Books, Guelph, Canada.

Lemke, S.L., Grant, P.G. and Phillips, T.D. (1998). Adsorption of zearalenone by organophilic montmorillonite clay. *Journal of Agricultural Food Chemistry* 46:3789–3796.

Lemke, S.L., Mayura, K., Reeves, W.R., Wang, N., Fickey, C. and Phillips, T.D. (2001a). Investigation of organophilic montmorillonite clay inclusion in zearalenone contaminated diets using the mouse uterine weight bioassay. *Journal of Toxicology and Environmental Health* 62:243–258.

Lemke, S.L., Ottinger, S.E., Mayura, K., Ake, C.L., Pimpukdee, K., Wang, N. and Phillips, T.D. (2001b). Development of a multi-tiered approach to the in vitro prescreening of clay-based enterosorbents. *Animal Feed Science and Technology* 93(1–2):17–29.

Li, M.H., Raverty, S.A. and Robinson, E.H. (1994). Effects of dietary mycotoxins produced by the mold *Fusarium moniliforme* on channel catfish *Ictalurus punctatus*. *Journal of the World Aquaculture Society* 25:512–516.

Lieschke, G. and Trede, N.S. (2009). Fish immunology. *Current Biology* 19(16):678–682.

Lightner, D.V. (ed.) (1996). *A Handbook of Shrimp Pathology and Diagnostic Procedures for Diseases of Cultured Penaeid Shrimp*. World Aquaculture Society, Baton Rouge, Louisiana.

Lindahl, J.F., Kagera, I.N. and Grace, D. (2018). Aflatoxin M_1 levels in different marketed milk products in Nairobi, Kenya. *Mycotoxin Research* 34(4):289–295.

Logrieco, A.F.J., Miller, D., Eskola, M., Krska, R., Ayalew, A., Bandyopadhyay, R., Battilani, P., Bhatnagar, D., Chulze, D., De Saeger, S., Li, P., Perrone, G., Poapolathep, A., Rahayu. E.S., Shephard, G.S., Stepman, F., Zhang, H. and Leslie, J.F. (2018). The mycotox charter: Increasing awareness of, and concerted action for, minimizing mycotoxin exposure worldwide. *Toxins* 10:149.

Loker, E.S., Adema, C.M., Ming-Zhang, S., Kepler, T. (2004). Invertebrate immune systems – not homogeneous, not simple, not well understood. *Immunology Reviews* 198:10–24.

Los, A., Ziuzina, D. and Bourke, P. (2018). Current and future technologies for microbiological decontamination of cereal grains. *Journal of Food Science* 83:1484–1493.

Lovell, R.T. (1992). Mycotoxins: Hazardous to farmed fish. *Feed International*: 24–28.

Lukuyu, B., Okike, I., Duncan, A., Beveridge, M. and Blümmel, M. (2014). Use of cassava in livestock and aquaculture feeding programs, Nairobi, Kenya. *ILRI Discussion Paper 25*.

Lumlertdacha, S. and Lovell, R. (1995). Fumonisin-contaminated dietary corn reduced survival and antibody production by channel catfish challenged with *Edwardsiella ictaluri*. *Journal of Aquatic Animal Health* 7:1–8.

Luo, Y., Liu, X. and Li, J. (2018). Updating techniques on controlling mycotoxins – A review. *Food Control* 89:123–132.

Madhusudhanan, N., KavithaLakshmi, S.N., Radha Shanmugasundaram, K. and Shanmugasundaram, E.R.B. (2004). Oxidative damage to lipids and proteins induced by aflatoxin B_1 in fish (*Labeo rohita*) - protective role of Amrita Bindu. *Environmental Toxicology and Pharmacology* 17:73–77.

Magnadóttir B. (2006). Innate immunity of fish (overview). *Fish & Shellfish Immunology* 20:137–151.

Malekinejad, H., Maas-Bakker, R., Fink-Gremmels, J. (2006). Species differences in the hepatic biotransformation of zearalenone. *The Veterinary Journal* 172: 96–102.

Malir, F., Ostry, V., Pfohl-Leszkowicz, A., Malir, J. and Toman, J. (2016). Ochratoxin A: 50 years of research. *Toxins* 8:191. Doi:10.3390/toxins8070191.

Mallmann, C.A. and Dilkin, P. (2007). *Micotoxinas e micotoxicoses em suínos*. Sociedade Vicente Pallotti Editora, Santa Maria, Brazil.

Mankevičienė, A., Jablonskytė-Raščė, D. and Maikštėnienė, S. (2014). Occurrence of mycotoxins in spelt and common wheat grain and their products. *Food Additives and Contaminants: Part A* 31:132–138.

Mannaa, M. and Kim, K.D. (2017). Influence of temperature and water activity on deleterious fungi and mycotoxin production during grain storage. *Mycobiology* 45:240–254.

Manning, B.B. and Abbas, H.K. (2012). The effect of Fusarium mycotoxins deoxynivalenol, fumonisin, and moniliformin from contaminated moldy grains on aquaculture fish. *Toxin Reviews* 31:11–15.

Manning, B., Ulloa, R., Li, M., Robinson, E. and Rottinghaus, G. (2003). Ochratoxin A fed to channel catfish causes reduced growth and lesions of hepatopancreatic tissue. *Aquaculture* 219:739–750.

Manning, B.B., Li, M.H. and Robinson, E.H. (2005a). Aflatoxins from mouldy corn cause no reductions in channel catfish Ictalurus punctatus performance. *Journal of the World Aquaculture Society* 36:59–67.

Manning, B.B., Terhune, J.S., Li, M.H., Robinson, E.H., Wise, D.J. and Rottinghau, G.E. (2005b). Exposure to feedborne mycotoxins T-2 Toxin or Ochratoxin A causes increased mortality of channel catfish challenged with *Edwardsiella ictaluri*. *Journal of Aquatic Animal Health* 17:147–152.

Manning, B.B., Abbas, H.K., Wise, D.J. and Greenway, T. (2014). The effect of feeding diets containing deoxynivalenol contaminated corn on channel catfish (*Ictalurus punctatus*) challenged with *Edwardsiella ictaluri*. *Aquaculture Research* 45:1782–1786.

Marin S., Ramos A. J., Cano-Sancho G. and Sanchis V. (2013). Mycotoxins: Occurrence, toxicology, and exposure assessment. *Food and Chemical Toxicology* 60:218–237.

Marroquín-Cardona, A., Deng, Y., Taylor, J.F., Hallmark, C.T., Johnson, N.M. and Phillips, T.D. (2009). In vitro and in vivo characterization of mycotoxin-binding additives used for animal feeds in Mexico. *Food Additives and Contaminants* 26:733–743.

Marth, E.H. (1992). Mycotoxin: Production and control. *Food Laboratory News*, pp.35–51.

Masching, S., Naehrer, K., Schwartz-Zimmermann, H.E., Sarandan, M., Schaumberger, S., Dohnal, I., Nagl, V. and Schatzmayr, D. (2016). Gastrointestinal degradation of fumonisin B_1 by carboxylesterase fumonisin D prevents fumonisin induced alteration of sphingolipid metabolism in turkey and swine. *Toxins* 8:84.

Massoud, R., Cruz, A. and Darani, K.K. (2018). Ochratoxin A: From safety aspects to prevention and remediation strategies. *Current Nutrition and Food Science* 14:11–16.

Matejova, I., Modra, H., Blahova, J., Franc, A., Fictum, P., Sevcikova, M. and Svobodova, Z. (2014). The effect of mycotoxin deoxynivalenol on haematological and biochemical indicators and histopathological changes in rainbow trout (*Oncorhynchus mykiss*). *BioMed Research International* 2014:310680.

Matejova, I., Vicenova, M., Vojtek, L., Kudlackova, H., Nedbalcova, K., Faldyna, M., Sisperova, E., Modra, H. and Svobodova, Z. (2015). Effect of the mycotoxin deoxynivalenol on the immune responses of rainbow trout (*Oncorhynchus mykiss*). *Veterinarni Medicina* 60:515–521.

Mayer, E., Novak, B., Springler, A., Schwartz-Zimmermann, H.E., Nagl, V., Reisinger, N., Hessenberger, S. and Schatzmayr, G. (2017). Effects of deoxynivalenol (DON) and its microbial biotransformation product deepoxy-deoxynivalenol (DOM-1) on a trout, pig, mouse, and human cell line. *Mycotoxin Research* 3:297–308.

Mbahinzireki, G.B., Dabrowski, K., Lee, K.J., El-Saidy, D. and Wisner, E.R. (2001). Growth, feed utilization and body composition of tilapia (*Oreochromis Spp.*) fed cottonseed meal-based diets in a recirculating system. *Aquaculture Nutrition* 7:189–200.

McCormick, S.D., Shrimpton, J.M., Carey, J.B., O'Dea, M.F., Sloan, K.E., Moriyama, S. and Björnsson, B.T. (1998). Repeated acute stress reduces growth rate of Atlantic salmon parr and alters plas levels of growth hormone, insulin-like growth I and cortisol. *Aquaculture* 168:221–235.

McKean, C., Tang, L., Billam, M., Tang, M., Theodorakis, C.W., Kendall, R.J. and Wang, J.S. (2006a). Comparative acute and combinative toxicity of aflatoxin B and T-2 toxin in animals and immortalized human cell lines. *Journal of Applied Toxicology* 26:139–147.

McKean, C., Tang, L., Tang, M., Billam, M., Wang, Z., Theodorakis, C.W., Kendall, R.J. and Wang, J.S. (2006b). Comparative acute and combinative toxicity of aflatoxin B_1 and fumonisin B_1 in animals and human cells. *Food and Chemical Toxicology* 44:868–876.

Meredith, F.I., Riley, R.T., Bacon, C.W., Williams, D.E. and Carlson, D.B. (1998). Extraction, quantification, and biological availability of fumonisin B_1 incorporated into the Oregon test diet and fed to rainbow trout. *Journal of Food Protection* 61:1034–1038.

Merrill Jr., A.H., Sullards, M.C., Wang, E., Voss, K.A. and Riley, R.T. (2001). Sphingolipid metabolism: Roles in signal transduction and disruption by fumonisins. *Environmental Health Perspectives* 109:283–289.

Meyer, K., Mohr, K., Bauer, J., Horn, P. and Kovács, M. (2003). Residue formation of fumonisin B_1 in porcine tissues. *Food Additives and Contaminants* 20:639–647.

Michelin, E.C., Massocco, M.M., Godoy, S.H.S., Baldin, J.C., Yasui, G.S., Lima, C.G., Rottinghaus, G.E., Sousa, R.L.M. and Fernandes, A.M. (2017). Carry-over of aflatoxins from feed to lambari fish (*Astyanax altiparanae*) tissues. *Food Additives and Contaminants: Part A* 34:265–272.

Miller, J.D. and Trenholm, H.L. (eds) (1994). *Mycotoxins in Grain: Compounds other than Aflatoxins*. Eagan Press, St. Paul, Minnesota.

Minervini, F. and Dell'Aquila, M.E. (2008). Zearalenone and reproductive function in farm animals. *International Journal of Molecular Sciences* 9:2570–2584.

Molnar, O., Schatzmayr, G., Fuchs, E. and Prillinger, H.J. (2004). Trichosporon mycotinivorans sp. nov., a new yeast species useful in biological detoxication of various mycotoxins. *Applied and Systematic Microbiology* 27:661-671.

Montes-Belmont, R., Méndez-Ramírez, I. and Flores-Moctezuma, E. (2002). Relationship between sorghum ergot, sowing dates, and climatic variables in Morelos, Mexico. *Crop Protection* 21:899–905.

Nagl, V., Woechtl, B., Schwartz-Zimmermann, H.E., Hennig-Pauka. I., Moll, W.D., Adam, G., Berthiller, F. (2014). Metabolism of the masked mycotoxin deoxynivalenol-3-glucoside in pigs. *Toxicology Letters* 229:190–197.

Ngethe, S., Horsberg, T.E. and Ingebrigtsen, K. (1992). The disposition of 3H-aflatoxin B_1 in the rainbow trout (*Oncorhynchus mykiss*) after oral and intravenous administration. *Aquaculture* 108:323–332.

Ngethe, S., Horsberg, T.E., Mitema, E. and Ingebrigtsen, K. (1993). Species differences in hepatic concentration of orally administered ^{3}H-AFB$_1$ between rainbow trout (*Oncorhynchus mykiss*) and tilapia (*Oreochromis niloticus*). *Aquaculture* 114:355–358.

Ogunade, I.M., Martinez-Tuppia, C., Queiroz, O.C.M., Jiang, Y., Drouin, P., Wu, F., Vyas, D. and Adesogan, A.T. (2018). Silage review: Mycotoxins in silage: Occurrence, effects, prevention, and mitigation. *Journal of Dairy Science* 101:4034–4059.

Oliveira, S.T.L.d., Veneroni-Gouveia, G., Santurio, J.M. and Costa, M.M.d. (2013). *Aeromonas hydrophila* in tilapia (*Oreochromis niloticus*) after the intake of aflatoxins. *Arquivos do Instituto Biológico* 80:400–406.

Omodu, F.O., Adegboyega, C.O., Olawuyi, O.J., Chibundu, E., Michael, S., Rudolf, K. and Adedayo, O. (2013). Multi-mycotoxin contaminations in fish feeds from different agro-ecological zones in Nigeria. Tropentag 2013, International Research on Food Security, Natural Resource Management and Rural Development.

Ostland V.E., Ferguson, H.W., Armstrong, R.D., A. Asselin and Hall., R. (1987). Granulomatous peritonitis in fish associated with Fusarium solani. *Veterinary Record* 121:595–596.

Ostrowski-Meissner, H.T. (1984). Effect of contamination of foods by *Aspergillus flavus* on the nutritive value of protein. *Journal of the Science of Food and Agriculture* 35:47–58.

Ostrowski-Meissner, H.T., LeaMaster, B.R., and Walsh, W.A. (1995). Sensitivity of the Pacific white shrimp, *Penaeus vannamei*, to aflatoxin B_1. *Aquaculture* 131:155–164.

Ostry, V., Malir, F., Toman, J. and Grosse, Y. (2016). Mycotoxins as human carcinogens – the *IARC Monographs* classification. *Mycotoxin Research* 33(1):65–73.

Oswald, I.P., Desautels, C., Laffitte, J., Fournout, S., Peres, S., Odin, M., Le Bars, P., Le Bars, J. and Fairbrother, J.M. (2003). Mycotoxin fumonisin B_1 increases intestinal colonization by pathogenic *Escherichia coli* in pigs. *Applied and Environmental Microbiology* 69:5870–5874.

Ottinger, C.A. and Kaattari, S.L. (1998). Sensitivity of rainbow trout leucocytes to aflatoxin B_1. *Fish and Shellfish Immunology* 8:515–530.

Ottinger, C.A. and Kaattari, S.L. (2000). Long-term immune dysfunction in rainbow trout (*Oncorhynchus mykiss*) exposed as embryos to aflatoxin B_1. *Fish and Shellfish Immunology* 10:101–106.

Panaccione, D.G. (2005). Origins and significance of ergot alkaloid diversity in fungi. *FEMS Microbiology Letters* 251:9–17.

Park, D.L., Whitaker, T.B., Giesbrecht, F.G. and Njapau, H. (2000). Performance of three pneumatic probe samplers and four analytical methods used to estimate aflatoxins in bulk cottonseed. *Journal of the Association of Analytical Chemists International* 83(5):1247–1251.

Patriarca, A. and Pinto, V.F. (2017). Prevalence of mycotoxins in foods and decontamination. *Current Opinion in Food Science* 14:50–60.

Pavan Kumar, B., Ramudu, K.R. and Devi, B.C. (2014). Mini review on incorporation of cotton seed meal, an alternative to fish meal in aquaculture feeds. *International Journal of Biological Research* 2:7.

Peng, W.-X., Marchalb, J.L.M. and van der Poela, A.F.B. (2018). Strategies to prevent and reduce mycotoxins for compound feed manufacturing. *Animal Feed Science and Technology* 237:129–153.

Pepeljnjak, S., Petrinec, Z., Kovacic, S. and Segvic, M. (2003). Screening toxicity study in young carp (*Cyprinus carpio* L.) on feed amended with fumonisin B_1. *Mycopathologia* 156:139–145.

Pestka, J.J. (2007). Deoxynivalenol: toxicity, mechanisms and animal health risks. *Animal Feed Science and Technology* 137:283–298.

Pestka, J.J. (2010a). Deoxynivalenol: mechanisms of action, human exposure, and toxicological relevance. *Archives of Toxicology* 84:663–679.

Pestka, J.J. (2010b). Deoxynivalenol-induced proinflammatory gene expression: mechanisms and pathological sequelae. *Toxins* 2:1300–1317.

Pestka, J.J., Zhou, H.R., Moon, Y. and Chung, Y.J. (2004). Cellular and molecular mechanisms for immune modulation by deoxynivalenol and other trichothecenes: unravelling a paradox. *Toxicology Letters* 153:61–73.

Petrinec, Z., Pepeljnjak, S., Kovacic, S. and Krznaric, A. (2004). Fumonisin B_1 causes multiple lesions in common carp (*Cyprinus carpio*). *Deutsche tierärztliche Wochenschrift* 9:358–363.

Pfohl-Leszkowicz, A. and Manderville, R.A. (2007). Ochratoxin A: An overview on toxicity and carcinogenicity in animals and humans. *Molecular Nutrition and Food Research* 51:61–69.

Phillips, T.D., Lemke, S.L. and Grant, P.G. (2002). Characterization of clay-based enterosorbents for the prevention of aflatoxicosis. *Advances in Experimental Medicine and Biology* 504:157–171.

Pichler, E. and Rodrigues, I. (2017). Analyzing mycotoxin content in commodities/feeds. In: *Mycotoxins in Swine Production*. BIOMIN Edition, Erber AG, Austria.

Pierron, A., Mimoun, S., Murate, L.S., Loiseau, N., Lippi, Y., Bracarense, A.P., Liaubet, L., Schatzmayr, G., Berthiller, F., Moll, W.D., Oswald, I.P. (2016a). Intestinal toxicity of the masked mycotoxin deoxynivalenol-3-β-D-glucoside. *Archives of Toxicology* 90:2037–2046.

Pierron, A., Mimoun, S., Murate, L.S., Loiseau, N., Lippi, Y., Bracarense, A.P., Schatzmayr, G., He, J.W., Zhou, T., Moll, W.D. and Oswald, I.P. (2016b). Microbial biotransformation of DON: molecular basis for reduced toxicity. *Scientific Reports* 6:29105.

Pietsch, C. and Junge, R. (2016). Physiological responses of carp (*Cyprinus carpio* L.,) to dietary exposure to zearalenone (ZEN). *Comparative Biochemistry and Physiology Part C: Toxicology & Pharmacology* 188:52–59.

Pietsch, C., Kersten, S., Burkhardt-Holm, P., Valenta, H. and Dänicke, S. (2013). Occurrence of deoxynivalenol and zearalenone in commercial fish feed: An initial study. *Toxins* 5:184.

Pietsch, C., Michel, C., Kersten, S., Valenta, H., Dänicke, S. and Schulz, C. (2014). In vivo effects of deoxynivalenol (DON) on innate immune responses of carp (*Cyprinus carpio* L.,). *Food and Chemical Toxicology* 68:44–52.

Pietsch, C., Kersten, S., Valenta, H., Dänicke, S., Schulz, C., Burkhardt-Holm, P. and Junge, R. (2015). Effects of dietary exposure to zearalenone (ZEN) on carp (*Cyprinus carpio* L.). *Toxins* 7:3465.

Plumb, J.A., Horowitz, S.A. and Rogers, W.A. (1986). Feed-related anemia in cultured channel catfish (*Ictalurus punctatus*). *Aquaculture* 51:175–179.

Poppenberger, B., Berthiller, F., Lucyshyn, D., Sieberer, T., Schuhmacher, R., Krska, R., Kuchler, K., Glössl, J., Luschnig, C. and Adam, G. (2003). Detoxification of the Fusarium mycotoxin deoxynivalenol by a UDP-glucosyltransferase from *Arabidopsis thaliana*. *The Journal of Biological Chemistry* 278:47905–47914.

Pozzo, L., Cavallarin, L., Nucera, D., Antoniazzi, S. and Schiavone, A. (2010). A survey of ochratoxin A contamination in feeds and sera from organic and standard swine farms in northwest Italy. Journal of the Science of Food and Agriculture 90:1467–1472.

Prelusky, D.B., Miller, J.D. and Trenholm, H.L. (1996). Disposition of ^{14}C-derived residues in tissues of pigs fed radiolabelled fumonisin B_1. *Food Additives and Contaminants* 13:155–162.

Rajeev Raghavan, P., Zhu, X., Lei, W., Han, D., Yang, Y. and Xie, S. (2011). Low levels of Aflatoxin B_1 could cause mortalities in juvenile hybrid sturgeon, *Acipenser ruthenus* ♂ × *A. baeri* ♀. *Aquaculture Nutrition* 17:e39–e47.

Rana, K.J., Siriwardena, S. and Hasan, M.R. (2009). Impact of rising feed ingredient prices on aquafeeds and aquaculture production. *FAO Technical Paper No. 541*, Rome.

Rankin, M. and Grau, C. (2002). Agronomic considerations for molds and mycotoxins in corn silage. *Focus on Forage* 4:1–4.

Reid, L.M., Nicol, R.W., Ouellet, T., Savard, M., Miller, J.D., Young, J.C., Stewart, D.W. and Schaafsma, A.W. (1999). Interaction of *Fusarium graminearum* and *F. moniliforme* in maize ears: Disease progress, fungal biomass, and mycotoxin accumulation. *Phytopathology* 89(11):1028–1037.

Reus, A. (2017). Alltech: Global aquafeed production up 12 percent. Available at https://www.wattagnet.com/articles/29597-alltech-global-aquafeed-production-up-12-percent?v=preview.

Ribeiro, J.M.M., Cavaglieri, L.R., Fraga, M.E., Direito, G.M., Dalcero, A.M. and Rosa, C.A.R. (2006). Influence of water activity, temperature and time on mycotoxins production on barley rootlets. *Letters in Applied Microbiology* 42:179–184.

Riley, R.T., Enongene, E., Voss, K.A., Norred, W.P., Meredith, F.I., Sharma, R.P. et al. (2001). Sphingolipid perturbations as mechanisms for fumonisin carcinogenesis. *Environmental Health Perspectives* 109:301–308.

Riley, R.T. and Norred, W.P. (1996). Mechanisms of mycotoxicity. *The Mycota* 6:193–211.

Riley, R.T. and Voss, K.A. (2011). Developing mechanism-based and exposure biomarkers for mycotoxins in animals. In: de Saeger, S. (ed.), *Determining mycotoxins and mycotoxigenic fungi in food and feed*. Woodhead Publishing Limited, UK, pp.245–275.

Riley, R.T., An, N.H., Showker, J.L., Yoo, H.S., Norred, W.P., Chamberlain, W.J., Wang, E., Merrill Jr, A.H. and Motelin, G. (1993). Alteration of tissue and serum sphinganine to sphingosine ratio: an early biomarker of exposure to fumonisin-containing feeds in pigs. *Toxicology and Applied Pharmacology* 118:105–112.

Riley, R.T., Wang, E. and Merrill Jr, A.H. (1994). Liquid chromatographic determination of sphinganine and sphingosine: use of the free sphinganine-to-sphingosine ratio as a biomarker for consumption of fumonisins. *Journal of AOAC International* 77:533–540.

Riley, R.T., Wang, E., Schroeder, J.J., Smith, E.R., Plattner, R.D., Abbas, H., Yoo, H.S. and Merrill Jr, A.H. (1996), Evidence for disruption of sphingolipid metabolism as a contributing factor in the toxicity and carcinogenicity of fumonisins. *Natural Toxins* 4:3–15.

Rinchard J, Ciereszko A., Dabrowski, K. and Ottobre, J. (2000). Effects of gossypol on sperm viability and plasma sex steroid hormones in male sea lamprey, Petromyzon marinus. *Toxicology Letters* 99:111–189.

Ritieni, A., Monti, S.M., Moretti, A., Logrieco, A., Gallo, M., Ferracane, R. and Fogliano, V. (1999). Stability of fusaproliferin, a mycotoxin from Fusarium spp. *Journal of Science in Food Agriculture* 79:1676–1680.

Robinson, E.H. and Li, M.H. (1994). Use of plant proteins in catfish feeds: replacement of soybean meal with cottonseed meal and replacement of fish meal with soybean meal and cottonseed meal. *Journal of World Aquaculture Society* 25:271–276.

Robinson, E.H. and Tiersch., T.R. (1995). Effects of long-term feeding of cottonseed meal on growth, testis development and sperm motility of male channel catfish. *Journal of World Aquaculture Society* 26:426–431.

Rodrigues, I., Binder, E.M. and Schatzmayr, G. (2009). Microorganisms and their enzymes for detoxifying mycotoxins posing a risk to livestock animals. In: Appell, M., Kendra, D.F. and Trucksess, M.W. (eds), *Mycotoxin Prevention and Control in Agriculture*, ACS Symposium Series, Vol. 1031, pp.107–117. American Chemical Society, Washington, DC.

Rodrigues, I. and Naehrer, K. (2012). A three-year survey on the worldwide occurrence of mycotoxins in feedstuffs and feed. *Toxins* 4:663–675.

Rotter, B.A., Prelusky, D.B. and Pestka, J.J. (1996). Toxicology of deoxynivalenol (vomitoxin). *Journal of Toxicology and Environmental Health A* 48:1–34.

Ruby, D.S., Masood, A., and Fatmi, A. (2013). Effect of aflatoxin contaminated feed on growth and survival of fish *Labeo rohita* (Hamilton). *Current World Environment* 8.

Rusaini, R. and Owens, L. (2010). Insight into the lymphoid organ of penaeid prawns: a review. *Fish and Shellfish Immunology* 29(3):367–377.

Rychlik, M., Humpf, H.U., Marko, D., Dänicke, S., Mally, A., Berthiller, F., Klaffke, H. and Lorenz, N. (2014). Proposal of a comprehensive definition of modified and other forms of mycotoxins including "masked" mycotoxins. *Mycotoxin Research* 30:197–205.

Ryerse, I.A., Hooft, J.M., Bureau, D.P., Hayes, M.A. and Lumsden, J.S. (2015). Purified deoxynivalenol or feed restriction reduces mortality in rainbow trout, *Oncorhynchus* *mykiss* (Walbaum), with experimental bacterial coldwater disease but biologically relevant concentrations of deoxynivalenol do not impair the growth of *Flavobacterium psychrophilum*. *Journal of Fish Diseases* 38:809–819.

Sahoo, P.K. and Mukherjee, S.C. (2000). Immunosuppressive effects of aflatoxin B_1 in Indian major carp (*Labeo rohita*). *Comparative Immunology, Microbiology and Infectious Diseases* 24:143–149.

Sahoo, P.K. and Mukherjee, S.C. (2001a). Effect of dietary β-1,3 glucan on immune responses and disease resistance of healthy and aflatoxin B_1-induced immunocompromised rohu (*Labeo rohita* Hamilton). *Fish and Shellfish Immunology* 11:683–695.

Sahoo, P.K. and Mukherjee, S.C. (2001b). Immunosuppressive effects of aflatoxin B_1 in Indian major carp (*Labeo rohita*). *Comparative Immunology, Microbiology and Infectious Diseases* 24:143–149.

Samapundo, S., Devliehgere, F., De Meulenaer, B. and Debevere, J. (2005). Effect of water activity and temperature on growth and the relationship between fumonisin production and the radial growth of *Fusarium verticillioides* and fusarium proliferatum on corn. *Journal of Food Protection* 68:1054–1059.

Samarajeewa, U., Sen, A.C., Cohen, M.D. and Wei, C.I. (1990). Detoxification of aflatoxins in foods and feeds by physical and chemical methods. *Journal of Food Protection* 53:489–501.

Sanchis, V. (2004). Environmental conditions affecting mycotoxins. In: Magan, N. and Olsen, M. (eds): *Mycotoxins in food*. CRC Press, Boca Raton, Florida, pp.174–189.

Santacroce, M.P., Conversano, M.C., Casalino, E., Lai, O., Zizzadoro, C., Centoducati, G., Crescenzo, G. (2007). Aflatoxins in aquatic species: metabolism, toxicity and perspectives. *Reviews in Fish Biology and Fisheries* 18:99–130.

Santin, E. (2005). Mould growth and mycotoxin production. In: *The Mycotoxin Blue Book*. Nottingham University Press, Nottingham, pp.225–234.

Schaafsma, A.W., Tamburic-Illinic, L., Miller, J.D. and Hooker, D.C. (2001). Agronomic considerations for reducing deoxynivalenol in wheat grain. *Molecular and Physiological Pathology* 23:279–285.

Schatzmayr, G., Heidler, D., Fuchs, E., Mohnl, M., Täubel, M., Loibner, A.P., Braun, R. and Binder, E.M. (2003). Investigation of different yeast strains for the detoxification of ochratoxin A. *Mycotoxin Research* 19:124–128.

Schatzmayr, G., Täubel, M., Vekiru, E., Moll, M., Schatzmayr, D., Binder, E.M., Krska, R. and Loibner, A.P. (2006a). Detoxification of mycotoxins by biotransformation. In: Barug, D., Bhatnagar, D., van Egmond, H.P., van der Kamp, J.W., van Osenbruggen, W.A. and Visconti. A. (eds), *The Mycotoxin Factbook.* Academic Publisher, Wageningen, pp.363–375.

Schatzmayr, G., Zehner, F., Täubel, M., Schatzmayr, D., Klimitsch, A., Loibner, A.P. and Binder, E.M. (2006b). Microbiologicals for deactivating mycotoxins. *Molecular Nutrition and Food Research* 50:543–551.

Schatzmayr, G., Schatzmayr, D., Pichler, E., Täubel, M., Loibner, A.P. and Binder, E.M. (2006c). Novel approach to deactivate ochratoxin A. In: Njapau, H. (ed.), *Mycotoxins and Phytotoxins*. Academic Publishers, Wageningen, pp. 279–288.

Schwartz, P., Bucheli, T.D., Wettstein, F.E. and Burkhardt-Holm, P. (2013). Life-cycle exposure to the estrogenic mycotoxin zearalenone affects zebrafish (*Danio rerio*) development and reproduction. *Environmental Toxicology* 28:276–289.

Schwartz, P., Thorpe, K.L., Bucheli, T.D., Wettstein, F.E. and Burkhardt-Holm, P. (2010). Short-term exposure to the environmentally relevant estrogenic mycotoxin zearalenone impairs reproduction in fish. *Science of The Total Environment* 409:326–333.

Scott, P.M. (2009). Ergot alkaloids: extent of human and animal exposure. *World Mycotoxin Journal* 2:141–149.

Segvic Klaric, M. (2012). Adverse effects of combined mycotoxins. Arhiv za Higijenu Rada i Toksikologiju 63:519–530.

Sepahdari, A., Ebrahimzadeh Mosavi, H.A., Sharifpour, I., Khosravi, A., Motallebi, A.A., Mohseni, M., Kakoolaki, S., Pourali, H.R. and Hallajian, A. (2010). Effects of different dietary levels of AFB_1 on survival rate and growth factors of Beluga (*Huso huso*). *Iranian Journal of Fisheries Sciences* 9:141–150.

Shane, S.H. and Eaton, D.L. (1994). Economic issues associated with aflatoxins. In: Groopman, J.D. (ed.), *The Toxicology of Aflatoxins: Human Health, Veterinary, and Agricultural Significance*. Academic Press, San Diego, CA, pp. 513–527.

Sharma, R.P. and Salunkhe, D.K. (eds) (1991). *Mycotoxins and Phytotoxins*, Boca Raton, Florida.

Shiwei X., Zheng L., Wan M., Niu J., Yongjan Y. and Tian L. (2018). Effect of deoxynivalenol on growth performance, histological morphology, anti-oxidative ability and immune response of juvenile Pacific white shrimp, *Litopenaeus vannamei. Fish and Shellfish Immunology* 82:442–452.

Skaug, M.A. (1999). Analysis of Norwegian milk and infant formulas for ochratoxin A. *Food Additives and Contaminants* 16:75–78.

Smith, G.W., Constable, P.D., Tumbleson, M.E., Rottinghaus, G.E. and Haschek, W.M. (1999). Sequence of cardiovascular changes leading to pulmonary edema in swine fed fumonisincontaining culture material. *American Journal of Veterinary Research* 60:1292–1300.

Smith, J.E. and Henderson, R.S. (1991). *Mycotoxin in Animal Foods*. CRC Press, Boca Raton, Florida.

Smith, J.E., Lewis, C.W., Anderson, J.G. and Solomons, G.L. (1994). *Mycotoxins in Human Health*. Report EUR 16048 EN, European Commission, Directorate-General XII, Brussels.

Söderhäll, K. (ed.) (2010). Invertebrate immunity. *Advances in Experimental Medicine and Biology* 708.

Souheil, H., Vey, A., Thuet, P., and Trilles, J.-P. (1999). Pathogenic and toxic effects of *Fusarium oxysporum* (Schecht.) on survival and osmoregulatory capacity of Penaeus japonicus (Bate). *Aquaculture* 178:209–224.

Springler, A., Hessenberger, S., Reisinger, N., Kern, C., Nagl, V., Schatzmayr, G. and Mayer, E. (2017). Deoxynivalenol and its metabolite deepoxy-deoxynivalenol: multi-parameter analysis for the evaluation of cytotoxicity and cellular effects. *Mycotoxin Research* 33:25–37.

Streit, E., Schwab, C., Sulyok, M., Naehrer, K., Krska, R. and Schatzmayr, G. (2013). Multi-mycotoxin screening reveals the occurrence of 139 different secondary metabolites in feed and feed ingredients. *Toxins* 5:504–523.

Strickland, J.R., Looper, M.L., Matthews, J.C., Rosenkrans Jr, C.F., Flythe, M.D. and Brown, K.R. (2011). Board-invited review: St. Anthony's Fire in livestock: Causes, mechanisms, and potential solutions. *Journal of Animal Science* 89:1603–1626.

Supamattaya, K., Sukrakanchana, N., Boonyaratpalin, M., Schatzmayr, D. and Chittiwan, V. (2005). Effects of ochratoxin A and deoxynivalenol on growth performance and immuno-physiological parameters in black tiger shrimp (*Penaeus monodon*). *Journal of Science and Technology* 27(Suppl. 1):91–99.

Swanson, S.P., Nicoletti, J., Rood, H.D., Buck, W.B. and Cote, L.M. (1987). Metabolism of three trichothecene mycotoxins, T-2 Toxin, diacetoxyscirpenol and deoxynivalenol, by bovine rumen microorganisms. *Journal of Chromatography* 414:335–342.

Sweeney, M.J. and Dobson, A.D.W. (1998). Mycotoxin production by *Aspergillus, Fusarium* and *Penicillium* species. *International Journal of Food Microbiology* 43:141–158.

Tacon, A.G.J., Hasan, M.R. and Metian, M. (2011). Demand and supply of feed ingredients for farmed fish and crustaceans: trends and prospects. *FAO Fisheries and Aquaculture Technical Paper* No. 564:87.

Takahashi, N., Miranda, C.L., Henderson, M.C., Buhler, D.R., Williams, D.E. and Bailey, G.S. (1995). Inhibition of in vitro aflatoxin B_1-DNA binding in rainbow trout by CYP1A inhibitors: α-naphthoflavone, β-naphthoflavone and trout CYP1A1 peptide antibody. *Comparative Biochemistry and Physiology Part C: Pharmacology, Toxicology and Endocrinology* 110:273–280.

Tapia-Salazar, M., García-Pérez, O.D., Nieto-López, M.G., Villarreal-Cavazos, D.A., Gamboa-Delgado, J., Cruz-Suárez, L.E. and Ricque-Marie, D. (2017). Evaluating the efficacy of commercially available aflatoxin binders for decreasing the effects of aflatoxicosis on Pacific white shrimp *Litopenaeus vannamei. Hidrobiológica* 27:411–418.

Trigo-Stockli, D.M., Obaldo, L.G., Dominy, W.G. and Behnke, K.C. (2000). Utilization of deoxynivalenol-contaminated hard red winter wheat for shrimp feeds. *Journal of the World Aquaculture Society* 31:247–254.

Tuan, N.A., Grizzle, J.M., Lovell, R.T., Manning, B.B. and Rottinghaus, G.E. (2002). Growth and hepatic lesions of Nile tilapia (*Oreochromis niloticus*) fed diets containing aflatoxin B_1. *Aquaculture* 212:311–319.

Tuan, N.A., Manning, B.B., Lovell, R.T. and Rottinghaus, G.E. (2003). Responses of Nile tilapia (*Oreochromis niloticus*) fed diets containing different concentrations of moniliformin or fumonisin B_1. *Aquaculture* 217:515–528.

Tudzynski, P., Correia, T. and Keller, U. (2001). Biotechnology and genetics of ergot alkaloids. *Applied Microbiology and Biotechnology* 57:593–605.

Uribe C., Folch H., Enriquez R. and Moran G. (2011). Innate and adaptive immunity in teleost fish: a review. *Veterinarni Medicina* 56(10): 486–503.

Van de Braak, C.B.T. (2002). Haemocytic defence in black tiger shrimp (*Penaeus monodon*). PhD thesis, Wageningen University. Wageningen Institute of Animal Sciences, the Netherlands.

Varga, J., Rigó, K. and Téren, J. (2000). Degradation of ochratoxin A by Aspergillus species. *International Journal of Food Microbiology* 59:1–7.

Vasanthi, S. and Bhat, R.V. (1998). Mycotoxins in foods-occurrence, health & economic significance & food control measures. *Indian Journal of Medical Research* 108:212–224.

Vasquez, L., Alpuche, J., Maldonado, G., Agundis, C., Pereyra-Morales, A., Zenteno, E. (2008). Immunity mechanisms in crustaceans. *Innate Immunity* 15(3):179–188.

Vekiru, E., Fruhauf, S., Sahin, M., Ottner, F., Schatzmayr, G. and Krska, R. (2007). Investigation of various adsorbents for their ability to bind Aflatoxin B_1. *Mycotoxin Research* 23(1):27–33.

Vekiru, E., Hametner, C., Mitterbauer, R., Rechthaler, J., Adam, G., Schatzmayr, G., Krska, R., and Schuhmacher, R. (2010). Cleavage of zearalenone by *Trichosporon mycotoxinivorans* to a novel nonestrogenic metabolite. *Applied Environmental Microbiology* 76:2353–2359.

Vila-Donat, P., Marín, S., Sanchis, V. and Ramos, A.J. (2018). A review of the mycotoxin adsorbing agents, with an emphasis on their multi-binding capacity, for animal feed decontamination. *Food and Chemical Toxicology* 114:246–259.

Völkel, I., Schröer-Merker, E. and Czerny, C-P. (2011). The carry-over of mycotoxins in products of animal origin with special regard to its implications for the European food safety legislation. *Food and Nutrition Sciences* 2:852–867.

Voss, K.A., Smith, G.W., Haschek, W.M. (2007). Fumonisins: Toxicokinetics, mechanism of action and toxicity. *Animal Feed Science and Technology* 137:299–325.

Wang, J.-S., Kensler, T.W. and Groopman, J.D. (1998). Toxicants in food: fungal contaminants. In: Ioannides, C.E. (ed.), *Nutrition and Chemical Toxicity*. Wiley, Indianapolis, USA.

Wang, Y., Chai, T., Lu, G., Quan, C., Duan, H., Yao, M., Zucker, B.-A. and Schlenker, G. (2008). Simultaneous detection of airborne aflatoxin, ochratoxin and zearalenone in a poultry house by immunoaffinity clean-up and high-performance liquid chromatography. *Environmental Research* 107:139–144.

Wang, X., Wang, Y., Li, Y., Huang, M., Gao, Y., Xue, X., Zhang, H., Encarnação, P., Santos, G. and Gonçalves, R.A. (2016). Response of yellow catfish (*Pelteobagrus fulvidraco*) to different dietary concentrations of aflatoxin B_1 and evaluation of an aflatoxin binder in offsetting its negative effects. *Ciencias Marinas* 42:15–29.

Webb, P.A. (2003). Introduction to chemical adsorption analytical techniques and their applications to catalysis. Micromeritics Instrument Corporation, Technical Publications January 2003. http://www.micromeritics.com/

Weber, H.P. (1980). The molecular architecture of ergopeptines: A basis for biological interaction. In: Goldstein, M., Lieberman, A., Calne, D.B. and Thorner, M.O. (eds), *Ergot Compounds and Brain Function: Neuroendocrine and Neuropsychiatric Aspects*. Raven Press, New York, NY, pp.25–34.

Wegst, W. and Lingens, F. (1983). Bacterial degradation of ochratoxin A. *FEMS Microbiology Letters* 17:341–344.

Wilson, S.C., Brasel, T.L., Carriker, C.G., Fortenberry, G.D., Fogle, M.R., Martin, J.M., Wu, C., Andriychuk, L.A., Karunasena, E. and Straus, D.C. (2004). An investigation into techniques for cleaning mould-contaminated home contents. *Journal of Occupation, Environment and Hygiene* 1:442–447.

Wiseman, M.O., Price, R.L., Lightner, D.V. and Williams, R.R. (1982). Toxicity of aflatoxin B_1 to penaeid shrimp. *Applied and Environmental Microbiology* 44:1479–1481.

Wolf, H. and Jackson, E.W. (1963). Hepatomas in rainbow trout: descriptive and experimental epidemiology. *Science* 142:676–678.

Woodward, B., Young, L.G. and Lun, A.K. (1983). Vomitoxin in diets for rainbow trout (*Salmo gairdneri*). *Aquaculture* 35:93–101.

Woźny, M., Brzuzan, P., Łuczyński, M.K., Góra, M., Bidzińska, J. and Jurkiewicz, P. (2008). Effects of cyclopenta[c]phenanthrene and its derivatives on zona radiata protein, ERα, and CYP1A mRNA expression in liver of rainbow trout (*Oncorhynchus mykiss* Walbaum). *ChemicoBiological Interactactions* 174:60–68.

Woźny, M., Brzuzan, P., Gusiatin, M., Jakimiuk, E., Dobosz, S., Kuzminski, H. (2012). Influence of zearalenone on selected biochemical parameters in juvenile rainbow trout (*Oncorhynchus mykiss*). *Polish Journal of Veterinary Sciences* 15:221–225.

Woźny, M., Obremski, K., Jakimiuk, E., Gusiatin, M. and Brzuzan, P. (2013). Zearalenone contamination in rainbow trout farms in north-eastern Poland. *Aquaculture* 416–417:209–211.

Woźny, M., Dobosz, S., Obremski, K., Hliwa, P., Gomułka, P., Łakomiak, A., Różyński, R., Zalewski, T. and Brzuzan, P. (2015). Feed-borne exposure to zearalenone leads to advanced ovarian development and limited histopathological changes in the liver of premarket size rainbow trout. *Aquaculture* 448:71–81.

Woźny, M., Obremski, K., Zalewski, T., Mommens, M., Łakomiak, A. and Brzuzan, P. (2017). Transfer of zearalenone to the reproductive system of female rainbow trout spawners: A potential risk for aquaculture and fish consumers? *Food and Chemical Toxicology* 107:386–394.

Wu, Q., Dohnal, V., Huang, L., Ku, K., Wang, X., Chen, G. and Yuan, Z. (2011). Metabolic pathways of ochratoxin A. *Current Drug Metabolism* 12:1–10.

Yiannikouris, A., Andre, G., Poughon, L., Francois, J., Dussap, C.G., Jeminet, G., Bertin, G. and Jouany, J.P. (2006). Chemical and conformational study of the interactions involved in mycotoxin complexation with ß-D-Glucans. *Biomacromolecules* 7:1147–1155.

Yoshizawa, T., Takeda, H. and Ohi, T. (1983). Structure of a novel metabolite from deoxynivalenol, a trichothecene mycotoxin, in animals. *Agricultural Biology and Chemistry* 47:2133–2135.

Yu, Y., Niu, J., Yin, P., Liu, Y., Tian, L. and Xu, D. (2018). Detoxification and immunoprotection of Zn(II)-curcumin in juvenile Pacific white shrimp (*Litopenaeus vannamei*) feed with aflatoxin B_1. *Fish and Shellfish Immunology* 80:480–486.

Zain, M.E. (2011). Impact of mycotoxins on humans and animals. *Journal of Saudi Chemical Society* 15:129–144.

Zhao, W., Wang, L., Liu, M., Jiang, K., Wang, M., Yang, G., Qi, C. and Wang, B. (2017). Transcriptome, antioxidant enzyme activity and histopathology analysis of hepatopancreas from the white shrimp *Litopenaeus vannamei* fed with aflatoxin B_1(AFB_1). *Developmental and Comparatve Immunology* 74:69–81.

Zhao, W., Wang, L., Liu, M., Jiang, K., Xia S, Qi C and Wang, B. (2018). Analysis of the expression of metabolism-related genes and histopathology of the hepatopancreas of white shrimp *Litopenaeus vannamei* fed with aflatoxin B_1. *Aquaculture* 485:191–196.

Zinedine, A., Soriano, J.M., Moltó, J.C. and Mañes, J. (2007). Review on the toxicity, occurrence, metabolism, detoxification, regulations and intake of zearalenone: an oestrogenic mycotoxin. *Food Chemistry and Toxicology* 45:1–18.

Zou, J. and Secombes, C.J. (2016). The function of fish cytokines. *Biology* 5:23.

Zychowski, K.E., Hoffmann, A.R., Ly, H.J., Pohlenz, C., Buentello, A., Romoser, A., Gatlin, D.M. and Phillips, T.D. (2013a). The effect of Aflatoxin-B(1) on red drum (*Sciaenops ocellatus*) and assessment of dietary supplementation of NovaSil for the prevention of aflatoxicosis. *Toxins* 5:1555–1573.

Zychowski, K.E., Pohlenz, C., Mays, T., Romoser, A., Hume, M., Buentello, A., Gatlin Iii, D.M. and Phillips, T.D. (2013b). The effect of NovaSil dietary supplementation on the growth and health performance of Nile tilapia (*Oreochromis niloticus*) fed aflatoxin-B_1 contaminated feed. *Aquaculture* 376–379:117–123.

08

Index

BV - #0007 - 110920 - C184 - 240/170/10 [12] - CB - 9781789180596